61 Structure and Bonding

Cation Ordering and Electron Transfer

With Contributions by
C. Gleitzer J. B. Goodenough B. G. Hyde
M. O'Keeffe U. Weser

With 77 Figures and 14 Tables

Springer-Verlag
Berlin Heidelberg New York Tokyo

ISBN 3-540-15446-9 Springer-Verlag Berlin Heidelberg New York Tokyo
ISBN 0-387-15446-9 Springer Verlag New York Heidelberg Berlin Tokyo

Library of Congress Catalog Card Number 67-11280

Typesetting and printing: Schwetzinger Verlagsdruckerei GmbH, 6830 Schwetzingen, Germany
Bookbinding: J. Schäffer OHG, 6718 Grünstadt, Germany
2152/3140-543210

Table of Contents

Mixed-Valence Iron Oxides

C. Gleitzer[1] and J. B. Goodenough[2]

[1] Laboratoire de Chimie du Solide Minéral, Associé au CNRS, Université de Nancy I, Boite Postale 239, F-54506 Vandœuvre les Nancy, Cedex

[2] Inorganic Chemistry Laboratory, University of Oxford, South Parks Road, Oxford, OX1 3QR, Great Britain

Mixed valence has been considered with reference to three model systems: Fe_3O_4 (magnetite) is a representative of the ferrospinels and the iron oxides with spinel-related structures, $Fe_{1-\delta}O$ (wüstite) illustrates a system with random defects and clustering, $CaFeO_3$ (a perovskite) illustrates the consequences of strong Fe^{4+}–O–Fe^{4+} interactions. The influence of counter cations is particularly marked in systems like the spinels Fe_2SnO_4 and Fe_2MoO_4; more subtle effects are illustrated by a comparison of $CaFeO_3$ with $SrFeO_3$. The time τ_h for an electron transfer between cations of different valence has been considered relative to two times: the period $\omega_R^{-1} \simeq 10^{-12}$ s of an optical-mode lattice vibration that traps a mobile electron and the time τ_n for a Mössbauer nuclear excited state to decay to its ground state. The full range of electron-transfer times $\tau_h < 10^{-12}$ s to $\tau_h > 10^{-8}$ s has been found for both Fe–Fe interactions across shared site edges or faces and Fe–O–Fe interactions across shared site corners. Perturbations of the periodic potential of an iron array by lattice defects or substitutional ions is shown to increase τ_h. Moreover, electron-lattice interactions were found to induce cooperative Jahn-Teller distortions in some compounds and charge-density waves (CDW) in others. In the mixed-valent compound Fe_3O_4, electrostatic interactions between mobile ions combine with electron-lattice interactions to produce, with increasing temperature, transitions from ionic ordering (or a static CDW) in the temperature interval $120 < T \lesssim 250$ to correlated electron jumps below 120K to uncorrelated electron jumps above 250K and from no spin degeneracy below a magnetic-ordering temperature $T_c \simeq 850$K to a spin degeneracy above T_c for the minority-spin electrons. Thus the physical properties induced by mixed valence in iron oxides is seen to be complicated by the fact that the mobile electrons may be either localized or itinerant; in fact localized and itinerant 3d electrons may co-exist at the same ion.

Structure and Bonding 61

1 Introduction

1.1 Significance of Mixed-Valence Iron Oxides

1.1.1 Natural

Iron is a common element, so many minerals contain iron oxides as a major constituent, and many others contain iron as an impurity. The significance of this fact for the study of minerals has been reviewed by Burns et al.[1].

The iron oxides, especially the magnetite and hematite ores, are primary sources of industrial iron; these ores cross mixed-valence states in the blast-furnace – or any other reduction process – before reaching the metallic endpoint.

The original lodestone, eventually used in the first compasses, was a mixed-valent, ferrimagnetic magnetite ore. The direction of magnetization of such iron ores, cooled through their Curie temperatures at different geological periods, have played an important role in the development of modern geological theory, and the presence of minute magnetite particles in some biological organisms, from worms to homing pigeons, make possible an internal navigational system[2].

1.1.2 Technical

Iron oxides are to be found in a great variety of technical fields, and the presence of mixed-valence states proves detrimental to some applications and essential to others.

In the electronics industry, for example, the magnetic properties of the iron oxides are used extensively, but the dielectric losses associated with mixed iron valences represent a problem to be minimized by fabrication procedures.

Corrosion of iron and steel involves the formation of mixed-valence iron oxides; it is a problem of enormous economic importance that remains inadequately understood, and the formation of oxide layers on steel plates or bars during hot-rolling is an important problem in steel manufacture.

On the other hand, electrical conduction via mixed-valence states is essential for iron oxides used in electrochemical cells or for the mediation of catalytic reactions at iron oxides used as heterogeneous catalysts.

Whether an advantage or a drawback, the mixed-valence character of iron oxides has important technical consequences, and its role in charge transfer needs to be understood as deeply as possible.

1.1.3 Academic

Five oxidation states of iron, II–VI, are accessible in oxides. The principal oxidation states are II–IV; these all carry spontaneous atomic magnetic moments in oxides; the mixed-valence states of particular interest are associated with III/II and IV/III couples on crystallographically equivalent sites. The low-temperature disproportionation reaction $2\,Fe^{4+} \rightarrow Fe^{3+} + Fe^{5+}$ in $CaFeO_3$ is also of fundamental theoretical interest.

The mixed-valence iron oxides provide an experimental test-bed for studying the evolution of charge-transfer processes from the localized-electron to the itinerant-electron regimes. Moreover, it is possible to monitor the influence of the charge transfer on the interatomic magnetic coupling since the iron ions in oxides carry localized magnetic moments.

1.2 Definitions and Concepts

1.2.1 Nominal Valence

In the oxide MgO, the electrostatic Madelung energy gained by ionic bonding (point-charge model) is greater than the energy loss associated with the transfer of two electrons from each magnesium to each oxygen atom. The Madelung energy thus stabilizes the O^{2-}: $2p^6$ band relative to the Mg^{2+}: 3s band. Although covalent mixing reduces the effective charges on the ions, and hence the electrostatic stabilization of the O^{2-}: $2p^6$ valence band relative to the Mg^{2+}: 3s conduction band, covalent mixing raises the mean energy of the antibonding conduction band states and stabilizes the mean energy of the bonding valence-band states so as to leave the overall binding energy of the solid similar to that calculated from a point-charge model. In this case the nominal valence states are taken to be Mg^{2+} and O^{2-} even though the effective ionic charges are less than II and admixing of oxygen 2p character into the conduction band and magnesium 3s character into the valence band are recognized.

Similar considerations hold for transition-metal oxides, but the distribution of covalent mixing between 3d and 4s states of an iron ion varies with the Fe–O separation and the occupancy of the 3d orbitals. So long as the Fermi energy E_F lies above the top of the O^{2-}: $2p^6$ band, which is normally the case in transition metal oxides, the total number of 3d electrons per iron atom can normally be determined from the chemical composition and the assumption that oxygen is present as O^{2-} ions. Of course, the amount of oxygen 2p character mixed into the iron 3d orbitals increases significantly with decreasing Fe–O separation and increasing formal positive charge of the iron; nevertheless, it is customary to speak of a $3d^n$ manifold, where n signals the number of electrons in the primarily 3d states antibonding with respect to the Fe–O interactions. The fact that Fe–O covalent mixing does not change the symmetry character of the orbitals in question makes this notation convenient.

Electron-spin-resonance (ESR) experiments on iron-doped TiO_2[3)] have located the Fe^{3+}/Fe^{2+} couple nearly 3 eV above the octahedral-site Fe^{3+}: $3d^5$ level, which was found just above the O^{2-}: $2p^6$ band edge. Since the standard redox potential $E^0(Fe^{4+}/Fe^{3+})$ lies at the upper edge of the Fe^{3+}: $3d^5$ level and the smaller Fe–O separation associated with Fe^{4+} pushes the Fe^{4+}: $3d^5$ level above the $E^0(Fe^{4+}/Fe^{3+})$ level, it is apparent that the formal valence state Fe^{4+} can be accomodated in the octahedral interstice of an oxide. Stabilization of a nominal Fe^{5+} state requires significant covalent mixing with the σ-bonding orbitals of e_g symmetry and an attendant decrease in the Fe–O bond distance. This is possible because the octahedral site configuration is $t_{2g}^3e_g^0$ for Fe^{5+}. Stabilization of Fe^{6+} only occurs in tetrahedral coordination with configuration $e_g^2t_{2g}^0$, which allows strong covalent mixing with σ-bonding orbitals of t_{2g} symmetry and a reduced O–O repulsion to allow the formation of a short Fe–O bond. The shorter the Fe–O bond, the greater the proportion of covalent mixing into the 3d versus the 4s orbitals of larger radial extension.

1.2.2 Intraatomic Versus Interatomic Interactions

There are two limiting descriptions of electrons outside closed atomic cores: crystal-field theory and band theory[4)]. Crystal-field theory applies where the intraatomic interactions are much stronger than the interatomic interactions between like atoms; band theory applies where the interatomic interactions are stronger. For a single-valence compound, a measure of the strength of the intraatomic interactions is the energy separation U_n of successive ionization energies. The energies U_n are due to the electrostatic interactions e^2/r_{ij} between electrons of the same manifold. They decrease sensitively with increasing radial extension of the wave functions, and hence in a solid with increased covalent mixing, which extends the cation wave function out over the neighboring atoms. The large energy $U_5 \approx 3$ eV separating the $3d^6$ and $3d^5$ configurations at an octahedral-site Fe^{3+} ion in oxides is due to the particular stability of a localized, high-spin $3d^5$ configuration in which the e^2/r_{ij} interactions are minimized by keeping the electrons in different 3d orbitals. This is the stabilization responsible for Hund's highest-multiplicity rule for the free ions. The sixth electron does not enjoy the stabilization energy Δ_{ex} associated with occupancy of a different orbital, so U_5 is not reduced by Δ_{ex}. In all the other U_n, this energy reduction is present, so U_2, U_3 and U_4 are significantly smaller than U_5. However, the effective U_3 for an octahedral-site ion is increased by the cubic-field splitting Δ_c if high-spin, by Δ_{ex} if low-spin.

The strength of the interatomic interactions between like atoms is given by the resonance (or energy-transfer) integral $t_{ij} = b_{ij}f(\theta_{ij})$, where $b_{ij} \equiv (\psi_i, H'\psi_j) \simeq \varepsilon_{ij}(\psi_i, \psi_j)$ is the spin-independent resonance integral of conventional tight-binding band theory and $f(\theta_{ij})$ is a function of the angle θ_{ij} between localized spins at the two interacting cations. The perturbation H' of the one-electron potential at position R_j by the presence of a like cation at R_i gives rise to the one-electron energy ε_{ij}, and (ψ_i, ψ_j) is the overlap integral for wavefunctions centered at R_i and R_j. In the absence of localized spins, tight-binding theory gives a bandwidth $w \simeq 2bz$, where z is the number of nearest neighbors and b is the nearest-neighbor resonance integral.

The condition for localized electrons with spontaneous atomic moments is $U_n > w$; that for itinerant electrons with no spontaneous moment is $U_n < w$. The intermediate case $U_n \simeq w$ is of considerable theoretical interest. In the case of iron oxides, a $U_5 \simeq 3$ eV ensures a localized $3d^5$ majority-spin configuration at both Fe^{2+} and Fe^{3+} ions since the cubic-field splitting $\Delta_c < \Delta_{ex}$ is small enough to leave the ions in the high-spin state. However, localization of the minority-spin electron, particularly in the mixed Fe^{3+}/Fe^{2+} state, does not necessarily follow. Similarly, a much smaller U_4 will be seen to make $U_4 \simeq w$ for Fe^{4+} in the perovskites $A^{2+}FeO_3$.

Two types of mixed-valence compounds must be distinguished at the outset: those with mixed valency on energetically equivalent sites and those in which the different valences occupy inequivalent sites. Section 1.2.3 deals with the former case, Sect. 1.2.5 with the latter.

1.2.3 Classification of Mobile Charge Carriers in Mixed-Valence Compounds

With mixed-valence compounds, charge transfer does not require creation of a polar state, and a criterion for localized versus itinerant electrons depends not on the intraatomic energy defined by U_n, but on the ability of the structure to trap a mobile charge carrier with a local lattice deformation. The two limiting descriptions for mobile charge carriers in mixed-valence compounds are therefore "small-polaron" theory and "itinerant-electron" theory. We shall find below that we must also distinguish mobile charge carriers of „intermediate" character.

Itinerant electrons occupy band states, which are molecular orbitals for an entire crystal; they therefore belong equally to all like atoms on energetically equivalent sites and are described by band theory.

Small polarons are mobile electrons (or holes) of a mixed-valence configuration that either tunnel or hop from site to site (e.g. $Fe^{3+} + Fe^{2+} \rightarrow Fe^{2+} + Fe^{3+}$) in a time $\tau_h > \omega_R^{-1} \simeq 10^{-12}$ s, where ω_R is the frequency of the optical-mode vibration that captures the mobile charge carrier in a local lattice deformation. The potential of the charge carrier is stabilized at a deformed site relative to that at neighboring sites, and a thermal activation energy is required to equalize the potentials so that tunneling can occur. After a jump, a new local lattice deformation again traps the mobile charge carrier. Therefore, the charge carriers move in a diffusive mode with an activated mobility $\mu_e = (e_0D_0/kT) \cdot \exp(-\Delta G_m/kT)$ where $\Delta G_m = \Delta H_m - T\Delta S_m$ is the free energy required to equalize the potentials at neighboring sites, e_0 is the magnitude of the electronic charge, D_0 is the preexponential factor of the diffusion coefficient, and kT is the Boltzmann energy. This expression is simply that due to Einstein for the drift mobility: $u_e = e_0D/kT$. Because the local deformation moves with the electron (or hole), the charge carriers are said to be "dressed" by their local deformation. Small polarons are thus defined as localized, but mobile charge carriers that are dressed in a local deformation and have an activated mobility.

Intermediate electrons are defined in this review as electrons having a high-temperature mobility given by diffusion theory, as in the case of small polarons, but with a motional enthalpy $\Delta H_m < kT$ so that $u_e \simeq eD_0/kT$ is unactivated. At low temperatures, electron-phonon interactions create small polarons if the concentration c of mobile charge carriers on the energetically equivalent sites is so small ($c \ll 1/2$) that the electrostatic coupling between mobile electrons is unimportant; but they create an ordering of the charge carriers if the concentrations are large enough to make the electrostatic forces important, e.g. $c = 1/2$. (The case $c \ll 1/2$ is generally realized only by doping with solute impurities or native detects, which perturb the electronic potential.) At intermediate temperatures, mobile-charge ordering is short-range and dynamic, at lowest temperatures it becomes static and induces changes in the lattice symmetry and/or periodicity. The mobile electrons may order on cationic sites to give identifyable valence states, or they may order in molecular-cluster states with fast or slow hopping between cations of a cluster. The Seebeck coefficient is given by small-polaron statistics at high temperatures; at intermediate temperatures the electron-phonon correlations change these statistics by allowing a mobile electron access to all nearest neighbors. Intermediate electrons are illustrated and treated more fully below in the section on ferrospinels, particularly in the discussion on magnetite.

Analogy with chemical classification of molecular compounds. Chemists are perhaps more familiar with the Robin-Day[5] classification of mixed-valence compounds, which was later refined by Mayoh and Day[6].

This classification distinguishes two degrees of "mixing" of two different configurations of a mixed-valence dimer: $M_A^{n+} - M_B^{(n+1)+}$ and $M_A^{(n+1)+} - M_B^{n+}$, where A and B refer to the two dimeric sites. Class I represents the limit of small mixing and is analogous to a small polaron with a large hopping enthalpy ΔH_m in the extended solid; class III represents the limit of equal (or strong) mixing and is analogous to the itinerant-electron case in the solid; class II is the intermediate case in which ΔH_m is small.

Bipolarons are mobile electron pairs. They have been identified in a small temperature interval above the semiconductor-metal transition in Ti_4O_7[7]. In this case, they represent mobile electron-pair bonds in a mixed-valence compound, the electrons condensing into metal-metal homopolar bonds that are ordered at low temperatures, but become disordered and mobile at somewhat higher temperatures.

1.2.4 Fe–Fe Versus Fe–O–Fe Interactions

Two types of interactions between metal atoms must be distinguished in the iron oxides: direct Fe–Fe interactions across a shared site face or edge and Fe–O–Fe interactions across a shared site corner.

The Fe–Fe interactions vary sensitively with the ratio of the radial extension of the π-bonding orbitals to the Fe–Fe separation. This ratio is significantly greater for the minority-spin electron at an Fe^{2+} ion than for the majority-spin electrons because the minority-spin electron of a high-spin $3d^6$ configuration is more weakly bound to the iron nucleus (by some 3 eV for octahedral-site iron in rutile) than are the $3d^5$ majority-spin electrons of the same symmetry. Therefore Fe–Fe interactions are strongest for the Fe^{3+}/Fe^{2+} couple.

On the other hand, Fe–O–Fe interactions have overlap integrals that vary as the square of the covalent mixing parameters since only the anion admixture has a large enough radial extension for overlap over the top of the anion. The covalent-mixing parameter $\lambda \sim b^{ca}/\Delta E$ is proportional to the cation-anion resonance integral b^{ca} and inversely proportional to the difference in energy ΔE between the lowest unoccupied cation 3d orbital and the overlapping oxygen p orbital. This mixing is already large enough at an Fe^{3+} ion to make the Fe^{3+}–O–Fe^{3+} magnetic interaction strong; magnetic ordering temperatures in the iron oxides are well above room temperature. At an Fe^{4+} ion, ΔE is reduced by some 3 eV, so we may anticipate Fe^{4+}–O–Fe^{4+} interactions that are strong enough to make $w \gtrsim U_4$.

The cubic-field splitting Δ_c is a measure of the difference in the strength of the σ-bond and π-bond covalent mixing; it is therefore larger at an Fe^{4+} than at an Fe^{3+} ion. The intraatomic-exchange stabilization Δ_{ex} increases with the number of parallel-spin electrons on the ion, so Δ_{ex} is larger for Fe^{3+}: $3d^5$ than Fe^{4+}: $3d^4$ configurations. Therefore, a high-spin ($\Delta_c < \Delta_{ex}$) configuration at Fe^{3+} ions is compatible with a $\Delta_c \simeq \Delta_{ex}$ at Fe^{4+}

ions. If $\Delta_c < \Delta_{ex}$ for octahedral-site Fe^{4+} ions, a $w \simeq U_4$ may result in narrow-band σ-antibonding (σ*) electrons of e-orbital parentage due to Fe^{4+}–O–Fe^{4+} interactions; if $\Delta_c > \Delta_{ex}$, a low-spin $t_{2g}^4\sigma^0$ configuration is stabilized; and if $\Delta_c \approx \Delta_{ex}$, the presence of a narrow σ* band may tend to stabilize an intermediate-spin state.

The small $U_4 \simeq w$ at Fe^{4+} ions in oxides also makes possible the disproportionation reaction $2\,Fe^{4+} \rightarrow Fe^{3+} + Fe^{5+}$ in the perovskite $CaFeO_3$, where Fe^{4+}–O–Fe^{4+} interactions are dominant at higher temperatures.

In the case of mixed valence on the tetrahedral (A) sites of a ferrospinel, an important question is whether charge transfer proceeds via Fe_A–O–O–Fe_A transfer or via Fe_A–O–M_B–O–Fe_A transfer.

1.2.5 Energetically Inequivalent Iron Sites

If like atoms occupy energetically inequivalent lattice sites with different valence states, charge transfer between the two types of sites requires overcoming the site-preference energy barrier. In the ferrospinels, for example, the tetrahedral-site iron Fe_A and the octahedral-site iron Fe_B have different $3\,d^n$-state potential energies. Whether Fe^{3+} or Fe^{2+} has the stronger A-site preference depends upon the character of the counter cations present; whereas the Fe^{3+} ion has a clear A-site preference in Fe_3O_4, a change of sign occurs with increasing x in the system $Fe_{3-x}Cr_xO_4$ (see below). In Fe_3O_4, the electron-transfer reaction

$$Fe_B^{2+} + Fe_A^{3+} = Fe_B^{3+} + Fe_A^{2+}$$

is shifted to the left at low temperatures, the population of Fe_A^{2+} ions varying as $\exp(-\Delta G_{AB}/kT)$, where $\Delta G_{AB} = \Delta H_{AB} - T\Delta S_{AB}$ is the site-preference free energy.

1.2.6 Energetic Equivalence of Dissimilar Ions

In some cases, mixed valence states may appear simultaneously on two different atoms because of an overlap of the redox energies associated with the two atoms. In the ilmenite $FeTiO_3$, the valence states Fe^{2+} and Ti^{4+} are stabilized with an ordering of Fe and Ti on crystallographically inequivalent sites[8]; $FeVO_3$, on the other hand, has the corundum structure with Fe and V randomly distributed on equivalent sites with the valence states Fe^{3+} and V^{3+}[9]. This observation places the top of the Fe^{2+}: $3\,d^6$ band (level) below the bottom of the Ti^{4+}: $3\,d^1$ band, but above the top of the V^{4+}: $3\,d^2$ band, so charge transfers of the type

$$Fe^{2+} + Ti^{4+} = Fe^{3+} + Ti^{3+} \quad \text{or} \quad Fe^{3+} + V^{3+} = Fe^{2+} + V^{4+}$$

can be expected to require little energy. We shall see below that even less energy is involved in the electron-transfer reactions

$$Fe^{2+} + Sn^{4+} = Fe^{3+} + Sn^{3+} \quad \text{and} \quad Fe^{2+} + Mo^{4+} = Fe^{3+} + Mo^{3+}$$

in the ferrospinels

$$Fe[FeSn]O_4 \quad \text{and} \quad Fe[FeMo]O_4$$

1.3 Experimental

1.3.1 Synthetic Oxides Versus Minerals

Although there are many features common to synthetic oxides and minerals, fundamental studies of the charge-transfer processes in mixed-valence compounds can only be systematically carried out on synthetic oxides of controlled stoichiometry and impurity concentration. However, with the exception of Seebeck coefficients, transport measurements require single-crystal data if quantitative interpretations are to be made. Nevertheless, conductivity data for polycrystalline samples of cubic phases are useful if the samples are dense and care has been taken to eliminate any segregation of impurities into the grain boundaries.

Although a particular mineral property may be due to iron impurities, it is generally easier to show this dependance by preparing synthetic samples of variable iron concentration and extrapolating back to the impurity levels of interest in a particular mineral. For example, the iron impurities in blue beryl are at so low a concentration that only weak Mössbauer signals can be obtained.

Small monovalent cations such as Li^+ or Cu^+ may be mobile in iron oxides at relatively low temperatures. This fact can provide an opportunity to prepare unusual compounds at low temperatures by either chemical or electrochemical insertion or extraction of lithium or copper. However, care must be taken in high-temperature preparations to prevent, for example, loss of Li^+ as Li_2O at high temperatures or disproportionation at low temperatures – especially at grain boundaries – into lithium-rich and lithium-poor phases.

1.3.2 Oxygen Content

In addition to the usual chemical characterization, the properties of iron oxides vary sensitively with oxidation state, and it is commonly necessary to control the partial pressure of oxygen during preparation and during measurement at elevated temperatures.

1.3.3 Physical Characterization

In addition to X-ray and neutron-diffraction structural characterization, the physical properties of iron oxides have been studied by a wide variety of techniques. Most common are conventional transport, optical, dielectric, calorimetric and magnetic measurements. In addition, NMR and Mössbauer are widely used.

Because the iron ions carry a magnetic moment, the Hall data are difficult to interpret. The conventional theory of the Hall effect utilizes a spin-independent resonance (transfer-energy) integral, and an adequate theory incorporating a spin-dependent resonance integral needs to be developed for antiferromagnetic materials.

On the other hand, Mössbauer spectroscopy proves to be a particularly powerful tool for iron compounds. It allows an independent determination of the oxidation state at an iron ion through the *isomer shift* δ and provides information about the local point sym-

Table 1. Isomer-Shift Ranges (mm/s wrt Fe) in Iron Oxides

Coordination	Fe^{2+}	Fe^{3+}
IV	0.9–1.1	0.1–0.3
VI	1.1–1.2	0.2–0.4

metry at the iron through the *quadrupole splitting* Δ. In addition, the magnetic-ordering temperature and information about the strength of the internal molecular fields can be obtained from the *hyperfine splitting* occurring below the magnetic-ordering temperature.

In a mixed-valence system, the isomer shift is able to distinguish the jump time τ_h relative to the time $\tau_n \simeq 10^{-8}$ s for a Mössbauer nuclear excited state to decay to the ground state.

For $\tau_h \ll \tau_n$, the recorded spectrum is the weighted average of the individual oxida- within relatively narrow limits, can be associated with each valence state in different O^{2-}-ion coordinations. Table 1 gives the corresponding isomer shift δ in mm/s with respect to elemental iron as compiled by Gupta et al.[10] for high-spin Fe^{2+} and Fe^{3+} (mainly in spinels). An isomer shift $\delta \simeq 0.05$ mm/s wrt Fe given by Gallagher et al.[11] for Fe^{4+} is to be compared with the data of Table 3, Chap. 4.

For $\tau_h \ll \tau_n$, the recorded spectrum is the weighted average of the individual oxidation-state spectra. For the Fe^{3+}/Fe^{2+} couple, with a weighted mean valence state $\langle m \rangle$ corresponding to Fe(m), the isomer shifts for fourfold and sixfold coordination are, respectively,

$$\begin{aligned} \delta_{\langle m \rangle}(\mathrm{IV}) &\sim (2.6 - 0.8\,\langle m \rangle \pm 0.1)\ \mathrm{mm/s} \\ \delta_{\langle m \rangle}(\mathrm{VI}) &\sim (2.85 - 0.85\,\langle m \rangle \pm 0.1)\ \mathrm{mm/s} \end{aligned} \tag{1}$$

For $\tau_h \simeq \tau_n$, the observed spectra may change with temperature in response to the temperature dependence $\tau_h = \tau_0 \exp(\Delta H_m/kT)$ of the hopping time. Such a temperature dependence, although complicated in practice by the quadrupole splitting of iron spectra, allows in principle the determination of the hopping parameters τ_0 and ΔH_m, as shown by Berkooz et al. for Eu_3O_4[12] and by Brown et al.[13] for $Fe_3O(CH_3CO_2)_6(H_2O)_3$. It would be interesting to compare the value of ΔH_m obtained in this manner with that obtained from transport measurements.

Mixed valency with a $\tau_h \simeq \tau_n$ tends to broaden the linewidths of the Mössbauer spectra due to the superposition of several signals having hopping times in the range $10^{-8} < \tau_h < 10^{-7}$s. Line broadening with decreasing temperature T may indicate a $\tau_h < \tau_n$ that approaches 10^{-7}s at lower T.

Although attempts have been made to correlate the hyperfine field of a mixed-valence compound with the weighted mean of the Fermi contact-interaction field $H_s \simeq 220 < S_Z >$ kOe[10], the orbital and spin-moment contributions H_L and H_D, which may be comparable to H_s at Fe(II) ions, make any such correlations dubious[14].

2 Ferrospinels

2.1 Introduction

The ferrospinels are the most thoroughly investigated of the mixed-valence iron oxides for both theoretical and practical reasons. The cubic structure of Fig. 1, with a lattice parameter a_0, is quite simple. The anions are displaced from face-centered-cubic lattice positions by an amount $|u - (3/8)|\ a_0$ along the $\langle 111 \rangle$ directions in such a way as to preserve tetrahedral symmetry about cations occupying a subarray A of tetrahedral sites; a second cation subarray B consists of octahedral sites connected in three dimensions by shared edges. The molecular formulae are written $A[B_2]O_4$. The B-site configuration is such that the crystalline field at a B site has a trigonal component; an oxygen parameter $u > 3/8$ reduces the magnitude of this component. In oxide spinels, the separation between B sites is rather short: $d_{B-B} = a_0/2\sqrt{2} \simeq 3.0$ Å.

The physical properties of ferrospinels are sensitive to the iron-ion distribution on the two cation subarrays. Normally the Fe^{3+} ion has a stronger A-site preference than the Fe^{2+} ion, but the difference in site-preference energy depends upon the character of the counter cation, and in some spinels the Fe^{2+} valence state competes with the Fe^{3+} valence state for tetrahedral-site occupation.

Both Fe^{3+} and Fe^{2+} ions have a high-spin state in iron oxides. The majority-spin d^5 configuration is localized, but the minority-spin electron at an Fe^{2+} ion may not be. The character of the minority-spin electron is modulated by several factors:

(i) the magnitude of the Fe_B–Fe_B separation
(ii) the concentration of iron atoms on the B sites and their mean valence
(iii) the magnitude of the trigonal component of the B-site crystalline field
(iv) the character and distribution of the other cations

Therefore, the properties imparted by mixed valence in the ferrospinels are a complex function of the total chemical composition.

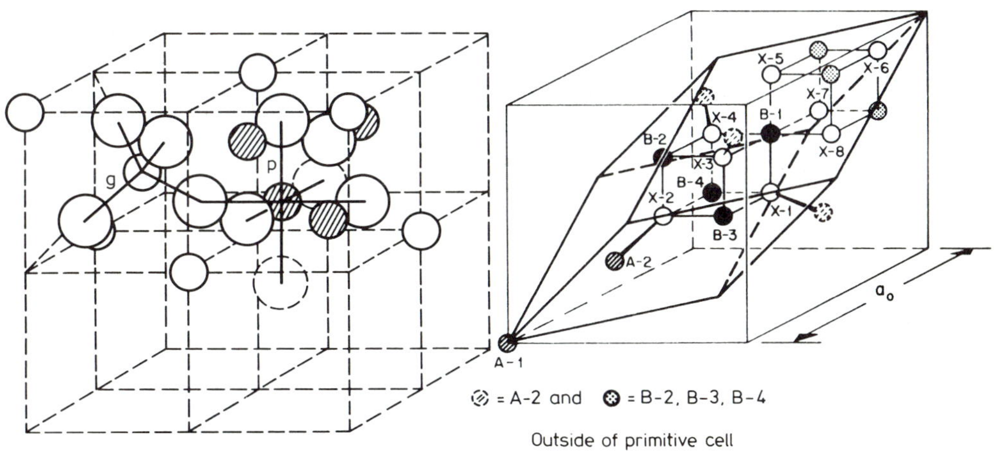

Fig. 1. Unit cell of spinel: (a) cubic and (b) primitive

The Néel model of ferrimagnetism in spinels is well known. It is based upon the assumption that the principal magnetic-exchange interaction is between the A and B sublattices and is negative. Consequently the magnetic order at low temperatures is collinear with A-site spins antiparallel to B-site spins, but with a net spontaneous magnetization. There are also exchange interactions between pairs of ions on the same subarray (A-A and B-B). They are usually much weaker; but, as Néel showed, they can profoundly influence the temperature dependence of the magnetization, which may display a maximum or invert its direction at a "compensation point". If the relative magnitudes of the A-A or B-B interactions become large enough, the magnetic ordering becomes complex; non-collinear spins replace antiparallel A-site and B-site spins. In tetragonal spinels, a triangular-spin (Yafet-Kittel) configuration may be stabilized; but in cubic spinels the triangular-spin configuration spirals about an axis perpendicular to the direction of spiral propagation[15)].

The crystalline anisotropy and magnetostriction are sensitive to the minority-spin electrons at Fe^{2+} ions, especially if this spin is localized and produces a cooperative Jahn-Teller distortion. The Fe^{3+} ion contribution to these properties arises primarily from dipole-dipole interactions.

Magnetite is chemically the simplest of the ferrospinels, but the physical properties it displays as a result of its mixed valency continue to provide a test-bed for our theories of charge transfer in the "intermediate regime". Therefore particular attention is given to magnetite as the basis for understanding the transport and magnetic properties exhibited by the more complex ferrospinels.

2.2 *Pure Magnetite*

2.2.1 *Stoichiometry*

Since the basic work of Darken and Gurry[16)], it is generally agreed that, under equilibrium conditions below 1000 °C, magnetite (Fe_3O_4) is nearly stoichiometric, the phase accepting a limited iron deficiency at higher temperatures. The iron-poor phase boundary at 1000 °C is about $Fe_{2.98}O_4$. In the iron-rich limit, Schmalzried et al.[17)] have identified a small proportion of interstitial iron at high temperature and low oxyen pressure. However, it must be stressed that the several reports[18–20)] of iron-rich magnetites $Fe_{3+x}O_4$, prepared by reducing iron oxides obtained by low-temperature synthesis, are dubious because of the presence of unsuspected bound water in the starting materials, which confuses the oxygen analysis by TGA[21)].

Under non-equilibrium conditions, it is possible to prepare the complete defect-spinel solid solution $Fe_{3-x}O_4$, $x < 0.33$[22)]. The iron vacancies are mainly on the B sites, so the end member γ-Fe_2O_3 may be written as $Fe^{3+}[Fe^{3+}_{5/3}\square_{1/3}]O_4$.

2.2.2 *The Verwey Transition*

At room temperature, magnetite has the cubic-spinel structure[23)], is a good electronic conductor, and exhibits a spontaneous magnetism.

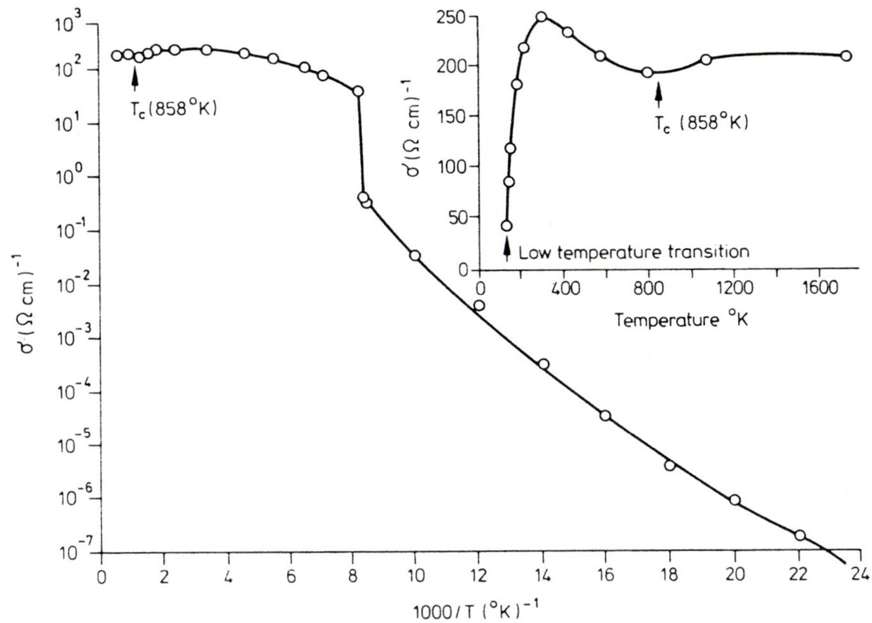

Fig. 2. Conductivity versus reciprocal temperature for stoichiometric Fe_3O_4, after[28)]

Scientific awareness of a low-temperature transition in magnetite began in 1929 with the observation[24)] of a Λ-type anomaly in the specific heat at about 120 K. The anomaly was typical of an order-disorder transition, but it was well below the magnetic-ordering temperature $T_c \simeq 850$ K[25)]. In 1931, Okamura[26)] observed an abrupt semiconductor-semiconductor transition near 120 K. The transition exhibits no thermal hysteresis, but the transition temperature is sensitive to the oxygen stoichiometry. More recent specific-heat measurements[27)] show the presence of two resolvable specific-heat peaks at the transition temperature; the lower-temperature peak near 110 K appears to be due to a spin reorientation.

Figure 2 illustrates conductivity versus temperature measurements made on single-crystal Fe_3O_4[28)].

In 1939, Verwey[29)] noted that the existence of a mixed valence on energetically equivalent sites permitted electronic conduction as the charge-transfer states

$$M^{m+} + M^{(m+1)+} = M^{(m+1)+} + M^{m+}$$

are then energetically equivalent. This equivalence holds even if small-polaron formation makes locally inequivalent the M^{m+} and $M^{(m+1)+}$ sites. In the absence of an external perturbation, the four crystallographically inequivalent B sites of spinel are energetically equivalent; they differ only in the direction of the trigonal component of the crystalline field. From this observation and the knowledge that Fe_3O_4 is a good electronic conductor at room temperature, Verwey made two deductions:

(i) Fe_3O_4 must be an inverse spinel with formula $Fe^{3+}[Fe^{3+}Fe^{2+}]O_4$

(ii) the transition at 120 K must represent an ordering of the mobile electrons into sites made inequivalent by a static lattice distortion to a lower-symmetry space group

Subsequently the transition has been called the Verwey transition and the transition temperature the Verwey temperature T_V. Verwey also guessed that, below T_V, the mobile electrons order as Fe^{2+} on [110] rows and Fe^{3+} on [1$\bar{1}$0] rows of B-site cations to produce a distortion to orthorhombic symmetry with lattice parameters approximately $(\sqrt{2} + \delta)a_0 \times (\sqrt{2} - \delta)a_0 \times a_0$, where δ is a small fraction and a_0 is the cubic lattice parameter. Although Bickford[30] was able to confirm that Fe_3O_4 is magnetically orthorhombic at temperatures $T < T_V$, it is now known that the low-temperature structure is in fact monoclinic with lattice parameters $\sqrt{2}a_0 \times \sqrt{2}a_0 \times 2a_0$ and that the electronic ordering is more complex than originally proposed by Verwey.

In 1948, Néel[31] postulated the existence of antiferromagnetic coupling between two cation sublattices in order to account for materials having a magnetic susceptibility

$$\chi = C/(T - \theta) \tag{2}$$

that obeyed a Curie-Weiss law, but had a negative Weiss constant ($\theta < 0$). If the two cation sublattices of such a material are identical, magnetic order leads to collinear-spin antiferromagnetism; if the A and B sublattices are different, as are the A and B sublattices of a spinel, magnetic order leads to collinear-spin ferrimagnetism with a net spontaneous magnetization per molecule

$$\bar{\mu}_{mol} = |\bar{\mu}_B - \bar{\mu}_A| \tag{3}$$

where $\bar{\mu}_B$ and $\bar{\mu}_A$ are the spontaneous magnetizations per molecule of the distinguishable sublattices. In the case of magnetite, a spin-only magnetization of $6\,\mu_B$ (μ_B = Bohr magneton) is predicted for the normal ionic distribution $Fe^{2+}[Fe_2^{3+}]O_4$ and $4\,\mu_B$ for the inverse distribution $Fe^{3+}[Fe^{3+}Fe^{2+}]O_4$. An observed spontaneous magnetization $\bar{\mu}_{mol}$ = $4.1\,\mu_B$[25] is consistent with the Verwey hypothesis, and direct confirmation of Néel ferrimagnetism has been obtained with neutron diffraction[32] and Mössbauer spectroscopy[33].

The existence of a collinear-spin ferrimagnetism below a $T_c \simeq 850$ K requires an antiferromagnetic exchange coupling between A and B sublattices that is negative and of large magnitude as well as intrasublattices couplings that are either small or positive. The strengths of these interactions are given by the exchange parameters J_{AB}, J_{AA}, and J_{BB} defined by the Heisenberg interatomic-exchange energy

$$H_{ij} = -\sum_{i \neq j} J_{ij}\mathbf{S}_i \cdot \mathbf{S}_j \tag{4}$$

From measurements of the spin-wave spectrum[34], it has been deduced that $J_{AB} \simeq -2.3$ meV, $J_{BB} \simeq +0.68$ meV, and J_{AA} is negligibly small. This observation is consistent with theoretical predictions for the signs and relative magnitudes of the superexchange interactions[35] as well as with the requirements for Néel ferrimagnetism. The interaction $J_{AB} < 0$ arises from Fe_A–O–Fe_B interactions between half-filled σ-bonding Fe_A-$t_2(\downarrow)$ and Fe_B-$e(\uparrow)$ orbitals; the interaction $J_{BB} > 0$ is due to the double-exchange coupling introduced by the mobile Fe_B-$a_1(\downarrow)$ electrons, which hop without changing their spin direction.

Below the Curie temperature T_c, the B-site moments are ferromagnetically coupled, and the mobile-electron spins on the B sites are coupled antiparallel to the B-site majority-spin direction, but parallel to the A-site majority-spin direction. Therefore, below T_c

the spin degeneracy of the mobile electron is removed and, since the majority-spin states are filled, charge transfer from B to A sites is inhibited. On the other hand, the number of minority-spin B-site electrons is only half the number of B-site iron atoms in Fe_3O_4, and the B sites are energetically equivalent; therefore the minority-spin electrons are mobile. However, whether to describe the electronic mobility by the itinerant-electron formula $u_e = e\tau/m^*$ or by the hopping (diffusion) formula $u_e = e_0D/kT$ must be determined experimentally. In order to do this, it is necessary first to know the electronic energy diagram for stoichiometric Fe_3O_4.

2.2.3 Energy Diagram

Construction of the semi-empirical room-temperature energy diagram of Fig. 3 for stoichiometric Fe_3O_4 begins with an estimate, from the optical spectrum[36)], of about 7 eV for the gap between the top of the filled $O^{2-}: 2p^6$ band and the empty Fe : 4s band. In TiO_2, the octahedral-site Fe^{3+}/Fe^{2+} level lies about 3 eV above the $O^{2-}: 2p^6$ band edge[3)]; in the spinel the top of the $O^{2-}: 2p^6$ band edge may be somewhat lower than in TiO_2 since it is σ-bonding rather than π-bonding to the cation array. Therefore the Fermi energy is placed about 3.8 eV above the top of the $O^{2-}: 2p^6$ band; it passes through the Fe_B^{3+}/Fe_B^{2+} redox band, which is half-filled. The empty Fe_A^{3+}/Fe_A^{2+} redox band lies an energy Δ above E_F; the smaller cubic-field splitting Δ_{cA} at a tetrahedral site ensures a $\Delta > 0$ while allowing the Fe_A-$t_2(\downarrow)$ electrons to be more stable than the Fe_B-$e(\uparrow)$ electrons. The

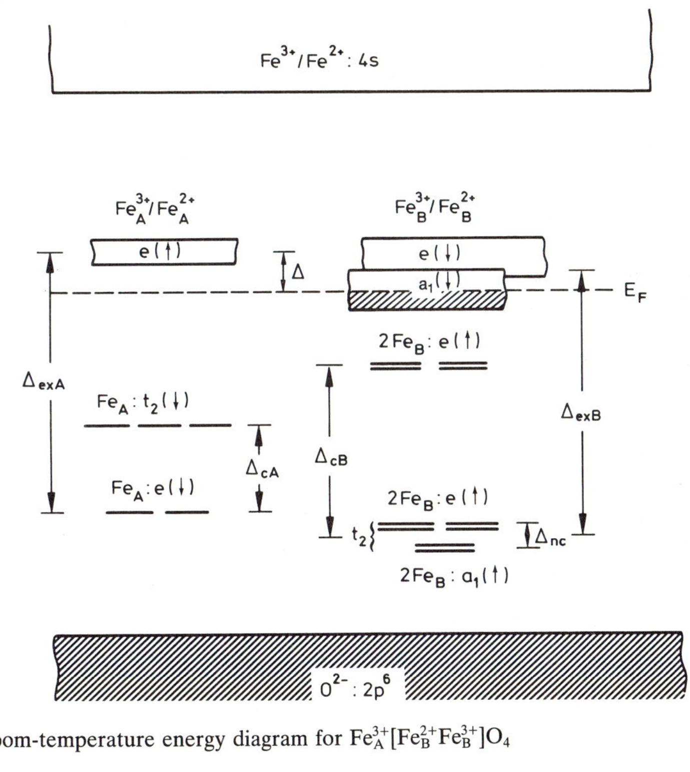

Fig. 3. Semi-empirical room-temperature energy diagram for $Fe_A^{3+}[Fe_B^{2+}Fe_B^{3+}]O_4$

arrows ↑ and ↓ represent spin directions relative to the net spin direction ↑ in the magnetically ordered state; and the partially occupied minority-spin Fe_B-$a_1(\downarrow)$ orbitals of the Fe_B^{3+}/Fe_B^{2+} redox couple have no spin degeneracy below T_c. Splitting by Δ_{nc} of the Fe_B-t_{2g} orbitals into a more stable a_1 orbital and twofold-degenerate e orbitals is induced by the trigonal component of the B-site crystalline field.

Polarized-spin UV photoelectron – spectroscopy measurements[37] confirm that the initial absorption is due to excitation of down-spin electrons, the Fe_B-$a_1(\downarrow)$ of the Fe_B^{3+}/Fe_B^{2+} redox couple, followed by the Fe_B-e(↑) ionization before the appearance of any Fe_A-$t_2(\downarrow)$ ionization. The data also place the Fe_A-$t_2(\downarrow)$ level about 1.5 eV below E_F. The optical data[36] seem to show an Fe_B-$a_1(\uparrow)$ level about 1 eV above the O^{2-}: $2p^6$ band edge, and a B-site cubic-field splitting $\Delta_{cB} \simeq 2$ eV would place the Fe_B-e(↑) level about 0.8 eV below E_F, which is consistent with the polarized-spin UV photoelectron-spectroscopy data[37]. An estimated A-site $\Delta_{cA} \simeq 1$ eV completes the diagram of Fig. 3. Note that with an octahedral-site Fe^{4+}/Fe^{3+} couple, the Fe_B-e(↑) level would be some 2 eV lower. The Fe_B-$a_1(\downarrow)$ electrons screen the other 3d(↑) electrons from the Fe_B nucleus, thereby raising their energies. In Fig. 3, it is assumed that the Fe_B-$a_1(\downarrow)$ electrons are sufficiently mobile that all of the Fe_B-3d(↑) electrons experience a similar shielding; no attempt is made to distinguish Fe_B^{3+} and Fe_B^{2+} species on the energy diagram.

At low temperatures ($T < T_V$) a crystal deformation from cubic to lower symmetry clearly splits in two the Fe_B^{3+}/Fe_B^{2+} band to stabilize occupied states at the expense of empty states. Although the particular ordering of Fe^{2+} and Fe^{3+} ions proposed by Verwey could produce such a splitting, his solution is clearly not unique. A similar splitting can be achieved, for example, with an itinerant-electron model for the mobile Fe_B electrons[38]. However, the character of a low-temperature distortion is generally different for localized versus itinerant electrons[4], which is why particular attention has been devoted to the structure and charge-density distribution of low-temperature ($T < T_V$) Fe_3O_4 during the last decade.

2.2.4 *High-Temperature ($T > T_V$) Magnetite*

Although the low-temperature charge distribution remains controversial, interpretation of the high-temperature data now appears possible.

It has been established[4] that the magnetic-ordering temperature goes through a maximum as the ratio of interatomic-to-intraatomic interactions (b/U) increases. Where this ratio is small, the electrons responsible for spontaneous magnetism may be described first by crystal-field theory, the interatomic superexchange interactions responsible for magnetic order entering only in perturbation theory. In Fe_3O_4, the Curie temperature T_c is determined primarily by the antiferromagnetic Fe_A–O–Fe_B superexchange interactions, which occur via overlap of the half-filled $t_{2A}(\downarrow)$ and $e_B(\uparrow)$ orbitals. A measured pressure dependence of the Curie temperature $dT_c/dP > 0$[39] shows that the ratio b/U for the Fe_A–O–Fe_B interactions is still less than that found for a maximum T_c, which defines the 3d electrons on the Fe_A^{3+} ions and the majority-spin 3d electrons on the Fe_B ions as localized in the sense that the superexchange second-order perturbation expansion converges. These electrons are so indicated in Fig. 3. However, the pressure dependence of T_c does not settle the status of the mobile majority-spin Fe_B-$a_1(\downarrow)$ electrons. First the character of the Fe_B-$a_1(\downarrow)$ electrons, like that of the Fe_B-$t_2(\uparrow)$ electrons, depends on the strength of the Fe_B–Fe_B interactions, not the Fe_B–O–Fe_A interactions. Second, although the Fe_B-$t_2(\uparrow)$ electrons are localized, the Fe_B-$a_1(\downarrow)$ electrons are less tightly bound to the Fe_B nuclei by $\Delta_{cB} \simeq 2$ eV, so their wavefunctions have a greater radial extension. Since the energy $b_{ij} \sim \varepsilon_{ij}(\psi_i\psi_j)$ contains an overlap integral that, for nearest-neighbor Fe_B–Fe_B interactions, varies sensitively with the Fe_B–Fe_B separation, it also increases sensitively with the radial extension of the overlapping orbitals of t_2 parentage. Therefore the parameter b for the Fe_B-$a_1(\downarrow)$ electrons is definitely larger than that for the localized Fe_B-$t_2(\uparrow)$ electrons. Third, the criterion for itinerant versus localized character for mobile electrons, where there is a mixed valency on energetically equivalent sites, is specified by the conditions for small-polaron formation, not by the condition for convergence of a superexchange perturbation expansion. Where the concentration of mobile charge carriers is large, as in Fe_3O_4, the polaron-polaron interactions make small-polaron formation depend on an ordering of the small polarons; therefore small-polaron formation is inhibited at higher temperatures. Finally, the magnetic ordering in Fe_3O_4 reduces the spin-disorder scattering within the ferromagnetic B-site subarray. Therefore, it is not possible to specify in advance the character of the Fe_B-$a_1(\downarrow)$ electrons even though the other 3d electrons in Fe_3O_4 are clearly localized.

In a cubic octahedral field, the minority-spin electron of a high-spin Fe^{2+} ion occupies a threefold-degenerate manifold of t_2 orbitals above the magnetic-ordering temperature. Ordering of the spins into a collinear configuration, as occurs in Fe_3O_4 below T_c, introduces an internal magnetic-exchange field, and hence a Zeeman energy. If the minority-spin electron is localized, as in $Fe_{1-\delta}O$, the Zeeman energy stabilizes a cooperative Jahn-Teller distortion that preserves the spin-orbit coupling[40]. The cooperativity arises from the long-range ordering of the spins; the spin-orbit coupling then introduces a giant "single-ion" magnetostriction and magnetic anisotropy. However, in Fe_3O_4 the trigonal-component of the B-site crystalline field lifts the t_2-manifold degeneracy, stabilizing the nondegenerate Fe_B-$a_1(\downarrow)$ state. Therefore the absence of a giant magnetostriction in Fe_3O_4 is not diagnostic of the character of the Fe_B-$a_1(\downarrow)$ electrons. Moreover, quenching of the first-order contribution of the orbital angular momentum means that the spontaneous magnetization should not be significantly changed by an ordering of the Fe_B-$a_1(\downarrow)$ electrons at T_V. Therefore the observation[41] that the change in spontaneous magnetization across T_V is only $\Delta M_s/M_s \simeq 0.1$ percent is not diagnostic either.

The electronic conductivity

$$\sigma = neu_e \tag{4}$$

is expected to have a constant density n of mobile charge carriers above T_V; thermal excitations to Fe_A sites do not alter n. If the mobile electrons are localized, the mobility is given by the Einstein expression for a diffusing species:

$$u_e = u_h = eD/kT\ , \quad D = D_0 \exp(\Delta S_m/k) \exp(-\Delta H_m/kT) \tag{5}$$

where e is the electronic charge, D is the diffusion coefficient, and kT is the Boltzmann temperature factor. The enthalpy ΔH_m for electron migration is the small-polaron energy of stabilization; in Fe_3O_4 the polaron-polaron interactions would reduce ΔH_m to a small value ($< kT$) in the disordered state at $T > T_V$. If, on the other hand, the mobile electrons are itinerant their mobility is given by

$$u_e = u_i = e\tau_s/m^* \tag{6}$$

where τ_s is the mean-free time between scattering events and m^* is the electron effective mass. From the Uncertainty Principle, the scattering time is

$$\tau_s > \hbar/w \tag{7}$$

where $\hbar$ is Planck's constant and the mobile-electron bandwidth is

$$w \simeq \hbar^2(\pi/a_0)^2/2\,m^* \tag{8}$$

Combining (6)–(8) gives

$$|u| > 2|e|a_0^2/\pi^2\hbar > 0.4\ \text{cm}^2/\text{Vs} \tag{9}$$

From Eq. (5), on the other hand, the upper limit for diffusional motion, which is found where $\Delta H_m = 0$, is

$$u_h = eD_0/kT\,, \quad D_0 = \gamma(1-c)l^2\nu_0 \tag{10}$$

where, from the theory of random walk, γ is a geometrical factor of order unity, $c = n/N$, N is the density of energetically equivalent sites available to the mobile electron, $l = a_0/2\sqrt{2}$ is a nearest-neighbor jump distance, and $\nu_0 \simeq 10^{12}$ Hz is an attempt frequency determined by optical-mode vibrations. At 300 K, the upper limit of the mobility is

$$|u_h| < |e|a_0^2\nu_0/kT \simeq 0.1\ \text{cm}^2/\text{Vs} \tag{11}$$

As shown in Fig. 2, the conductivity is nearly temperature-independent above 300 K except for a broad minimum at T_c characteristic of spin-disorder scattering and trapping of a fraction of the electrons in local (short-range) regions of magnetic order (magnetic polarons). A temperature-independent conductivity is characteristic of a constant n and an intermediate mobility having a mean-free distance between scattering events of only the nearest-neighbor distance l, as in diffusional motion, but with a $\Delta H_m < kT$. Since the mean-free distance between scattering events cannot be decreased further than one jump distance, it does not decrease further with increasing T in this limit, in contrast to a classical metal, so σ remains nearly temperature-independent. The magnitude of the room-temperature conductivity, 250 $\text{ohm}^{-1}\ \text{cm}^{-1}$, is also characteristic of this intermediate regime. Siemons[42)] has measured the Hall coefficient and Hall mobility on single-crystal Fe_3O_4 from room-temperature to below T_V. Above T_V, the Hall coefficient is essentially temperature-independent, indicative of a constant n, and the Hall mobility $|u_H| < 0.1$ cm^2/Vs, varies as $|u_H| \sim \exp(-T_0/T)$. These data are consistent with a diffusive motion, typical of small polarons, but with a $\Delta H_m < kT$ and a $\nu_0 = \nu_0(T)$.

A $\Delta H_m < kT$ has important consequences. As the temperature is lowered to where $\Delta H_m \simeq kT$, strong electron-phonon interactions must manifest themselves. Direct evidence for "mode softening" and strong electron-phonon coupling in the internal $T_V < T < 250$ K has been provided by measurements of the Mössbauer recoiless fraction and the X-ray Debye-Waller factor[43)] as well as of muon-spin rotation[44)]. Therefore, it would be reasonable to suspect that the semiconducting temperature dependence of the conductivity in the interval $T_V < T < 250$ K (see Fig. 2), which is also reflected in the temperature dependence of the mobility in this interval[42], is due to a short-range ordering of the mobile electrons in anticipation of the Verwey transition. Such a phenomenon, which produces a semiconductor-semiconductor transition at T_V, can be distinguished from the semiconductor-metal transitions found, for example, in VO_2[4]. Srinivasan and Srivastava[45] have developed a simple model of short-range cooperative electron-phonon coupling that is capable of quantitatively fitting the conductivity data in the temperature range $T_V < T < 250$ K with an expression

$$\sigma = (f_+ - f_-)ne^2l^2\nu_0/kT \qquad (12)$$

where the random-walk probability expression

$$f_+ - f_- = [\exp(\varepsilon_p/2\,kT) + \exp(-\,\varepsilon_p/2\,kT)]^{-2} \qquad (13)$$

contains the small-polaron stabilization energy $\varepsilon_p = 0.055$ eV associated with a dynamic correlation of Fe_B^{2+} and Fe_B^{3+} ions on alternate sites of a chain of edge-shared octahedra. Note that an identical expression would be obtained if the electron-lattice couplings condensed Fe_B^{2+}-Fe_B^{3+} pairs by alternating the Fe_B–Fe_B distances as short and long along the chain rather than Fe_B^{2+} and Fe_B^{3+} ions by altering the Fe_B–O distances. For a cation pair the polaron energy would refer to the stabilization of an $a_1(\downarrow)$ electron within a pair (molecular polaron). A third possibility is that the correlated electron transfer between ions is accompanied by a pairing such that the making and breaking of pairs is slow relative to the correlated electron transfer within a pair. In this latter case, mode softening may be associated with the pairing while the primary correlations are associated with electron transfers between ions. The three possible models are illustrated schematically in Fig. 4. In Eq. (12), the geometric factor γ is incorporated into the adjustable parameter $\nu_0 = 1.4 \times 10^{12}$ Hz. Since this is an electron-correlation model, the factor $f_+ - f_-$ modulates ν_0, which appears in the Hall mobility, and not n, which appears in the Hall coefficient. With this model, Srivastava[46] has also been able to account quantitatively for the observed changes in T_V with changing oxygen isotopes[47] and with changing pressure[48, 49]. He accomplishes this with the assumption that a minimum amplitude of the coupled vibrational mode is needed to sustain a dynamic correlation. The amplitude decreases with temperature to this minimum value at $T = T_V$, below which a static deformation is frozen in.

The Seebeck data of Fig. 5[50] provide information (1) on the thermal excitation of mobile electrons to the Fe_A sites, particularly above T_c, (2) the change in spin degeneracy at T_c, and (3) the influence of a $\Delta H_m < kT$ on the accessibility of nearest-neighbor sites.

The Seebeck coefficient

$$\alpha = (k/e)[(S_T/k) + (\varepsilon_F/kT)] \qquad (14)$$

contains a transport term and a statistical term. The lattice entropy S_T transported with electrons can generally be neglected in oxides. The statistical term contains the Fermi energy ε_F; in a mixed-valence system having a concentration c of mobile electrons on energetically equivalent sites, this term gives a Seebeck coefficient

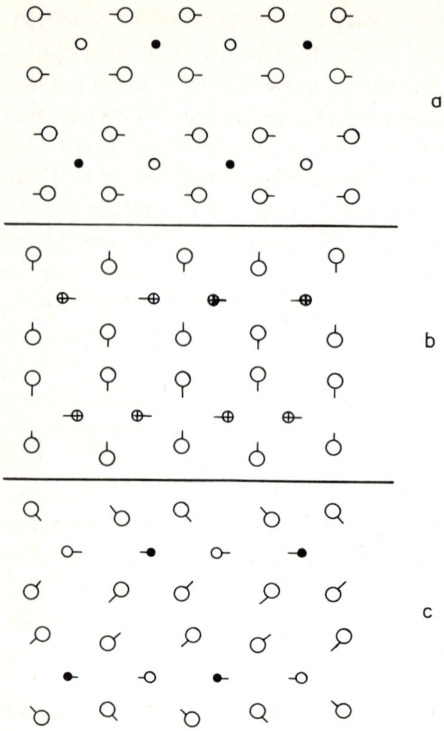

Fig. 4a–c. Models for electron hopping correlated by electrostatic electron-electron interactions plus strong electron-phonon interactions for a valence ratio $Fe^{2+}/Fe^{3+} \simeq 1$: (**a**) small polarons, (**b**) diatomic polarons, (**c**) small polaron coupled to slower (only one phase shown) dimerization

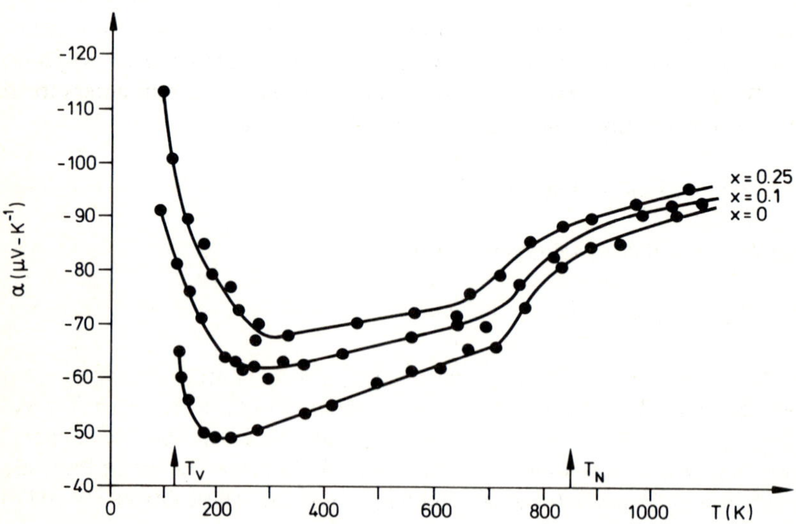

Fig. 5. Seebeck coefficient of $Fe_3O_{4-x}F_x$ versus temperature for $T > T_V$, after[50)]

$$\alpha \simeq -(k/e)\ln[\beta(1-c)/c] \tag{15}$$

where β is the spin degeneracy. At temperatures $T > T_c$, a $\beta = 2$ must be used; but well below T_c a $\beta = 1$ is appropriate. For a fixed β, this expression is temperature-independent so long as c is temperature-independent.

Three features of Fig. 4 are to be noted: (i) α is not temperature independent; (ii) at 300 K α is not small, as would be predicted from Eq. (15) with $c = 1/2$ and $\beta = 1$, and (iii) there is an abrupt, though smooth, decrease in the magnitude of α on lowering the temperature through T_c. Wu and Mason[51)] have attained a quantitative fit to the data for $T > T_c$, where $\beta = 2$, with Eq. (15) and a temperature-independent activation enthalpy and entropy in the energy $\Delta = \Delta H_{AB} - T\Delta S_{AB}$ of Fig. 3 for excitation of electrons from Fe_B to Fe_A sites. They assume that only the B sites contribute to α, so excitations to A sites lowers c in Eq. (15), thus raising the magnitude of α. A statistical distribution between A and B sites is reached at 1450 °C, where the cation-distribution equilibrium constant $K = \exp(-\Delta/kT)$ goes to unity. The abrupt decrease in the magnitude of α on lowering the temperature through T_c is clearly a reflection of the change in the spin-degeneracy parameter. However, α does not decrease to zero as T is lowered further; it levels off instead to a room-temperature value corresponding to

$$\alpha = -(k/e)\ln(1/c) \tag{16}$$

which suggests a transition to another transport regime.

Equation (15) is based on the assumption that the mobile electron can only jump from an Fe_B^{2+} ion to a near-neighbor Fe_B^{3+} ion, the factor $(1 - c)$ giving the probability that a near-neighbor is Fe_B^{3+} and not Fe_B^{2+}. Equation (16) implies that an electron can jump to any of its near-neighbor sites, and hence that electron-phonon interactions introduce correlated electron jumps that involve more than one electron. In the one-dimensional (1 D) theory of Srinavasan and Srivastava, each Fe_B^{2+} ions has only Fe_B^{3+} ions on the linear chain, and correlated motions along a chain give Eq. (16). In the 3 D problem of actual Fe_3O_4, it is not possible to organize the correlations so as to make every nearest neighbor of an Fe_B^{2+} ion an Fe_B^{3+} ion. Nevertheless, a $\Delta H_m < kT$ makes multiple-electron jumps competitive with one-electron jumps, and the dynamic correlations permit access to all nearest neighbors. For example, the two-electron charge transfer

$$Fe_B^{2+} + Fe_B^{2+} + Fe_B^{3+} \rightarrow Fe_B^{3+} + Fe_B^{2+} + Fe_B^{2+}$$

can be competitive with the one-electron transfer

$$Fe_B^{2+} + Fe_B^{3+} \rightarrow Fe_B^{3+} + Fe_B^{2+}$$

We thus conclude that electron-phonon interactions provide short-range correlations extending well above T_V that give a mobile electron access to any nearest-neighbor site with equal probability, so Eq. (16) replaces Eq. (15). Note that a change from hopping between ions to hopping between Fe_B^{3+}-Fe_B^{2+} pairs would alter the statistics as there are three times as many Fe_B–Fe_B bonds as there are Fe_B atoms, but a change from $c = 1/2$ to $c = 1/6$ is not compatible with the data. Therefore, it appears that at room-temperature and above the mobile electrons are moving between atoms with a $\Delta H_m < kT$. Below room-temperature, electron-phonon interactions may condense out either correlated Fe_B^{2+} and Fe_B^{3+} ions or correlated Fe_B^{2+}-Fe_B^{3+} pairs, but these correlations cannot change significantly the effective value of c until just above T_V, where the correlation length becomes longer range.

If a diffusional model with $\Delta H_m < kT$ is appropriate, then the time τ_h between electron hops must approach the period ω_R^{-1} of the optical mode vibrations that trap or correlate the electrons. With an $\omega_R^{-1} \simeq 10^{-12}$ s, the hopping time τ_h would be short relative to the time scale of Mössbauer spectroscopy, ca. 10^{-8} s. We can therefore anticipate an isomer shift for the octahedral-site iron that is midway between the values typical for Fe^{2+} ions and Fe^{3+} ions. From Table 1 and Eq. (1), we can predict a room-temperature isomer shift of $\delta^{2.5+} \sim 0.75$ mm/s wrt iron. Consistent with this prediction is the observation of only two Mössbauer signals at room-temperature, a sharp resonance (0.20 mm/s) readily assignable to A-site Fe^{3+} ions and a single, broader resonance (0.32 mm/s) for B-site $Fe^{2.5+}$ ions with an isomer shift $\delta \simeq 0.67$ mm/s[52)]. Broadening of the B-site resonance is due to quadrupole splittings associated with the trigonal component of the crystalline B-site fields; it does not constitute a proof of rapid electron transfer[53)]. However, the temperature independence of the line broadening between T_V and 300 K is consistent with a $\Delta H_m < kT$[53)]; and the single peak with an isomer shift typical of a mean valence 2.5 + rather than two resolved resonances with isomer shifts corresponding to octahedral-site Fe^{2+} and Fe^{3+} ions clearly demonstrates a $\tau_h < 10^{-8}$ s. In fact, a $\tau_h < 10^{-11}$ s has been established by selective-excitation double Mössbauer (SEDM) measurements[54)]; so the measured upper limit for τ_h approaches our anticipated value of about 10^{-12} s. The SEDM measurements also revealed an A-site relaxation at room-temperature; this observation is consistent with some excitation across Δ of Fig. 3, even at 300 K, that has a $\tau_h > 10^{-8}$ s. Moreover, some electron transfer to Fe_A^{3+} ions would reduce δ from $\delta^{2.5+} \simeq 0.75$ mm/s.

With an Fe_B^{2+}-Fe_B^{3+} pair model, the time to jump from pair to pair with a $\Delta H_m < kT$ would have to be about 10^{-12} s to be consistent with the Mössbauer data, thus requiring a $\tau_h < \omega_R^{-1}$ within a pair.

2.2.5 *Low-Temperature ($T < T_V$) Magnetite*

On lowering the temperature through T_V, a bandgap $E_g \simeq 0.1$ eV appears in the Fe_B-$a_1(\downarrow)$ conduction band of Fig. 3 at E_F. The Hall coefficient increases as $R_H \sim \exp(T_V/T)$, indicating that the charge-carrier density increases exponentially with T^{-1}, as in a normal semiconductor, and the Hall mobility increases from about 0.1 to 0.4 cm^2/Vs on lowering the temperature from $T_V \simeq 120$ K to 77 K[42)].

The low-temperature Seebeck coefficient exhibits the rather bizarre behavior shown in Fig. 6[55)]. As pointed out by Mott[56)], such a temperature dependence has a straightforward qualitative interpretation: with the opening of an energy gap E_g, the number of charge carriers is abruptly reduced, but a higher electron mobility than hole mobility gives a negative Seebeck coefficient. As the temperature is lowered further, p-type extrinsic conduction dominates; Fe_3O_4 samples tend to be slightly oxidized. However, the cation vacancies provide deep enough hole traps that, at lowest temperatures, impurity-band hopping takes over, and the Seebeck coefficient again becomes negative as a result of partial donor-impurity compensation.

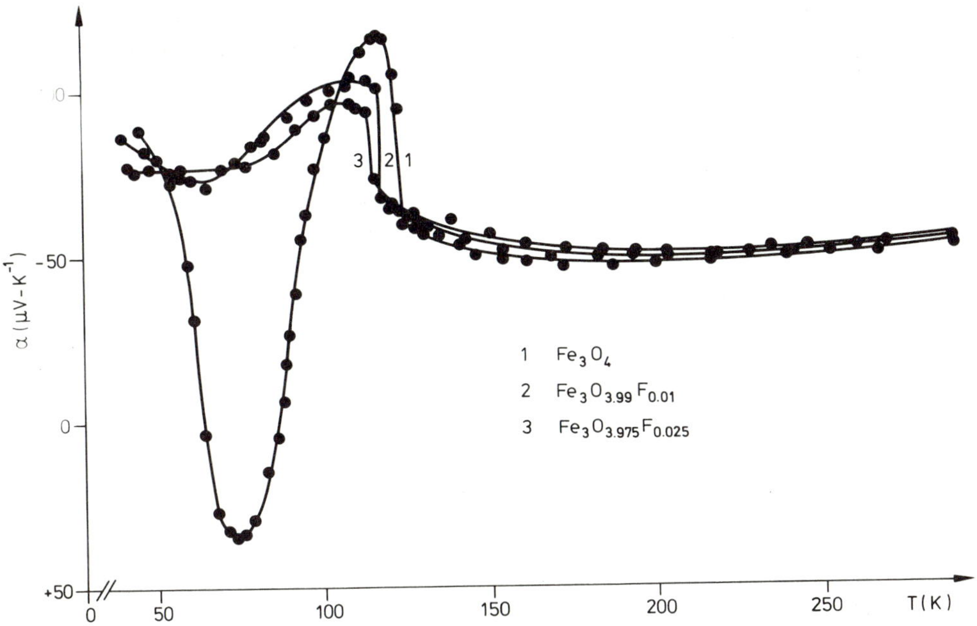

Fig. 6. Seebeck coefficient of $Fe_3O_{4-x}F_x$ versus temperature for $T < T_V$, after[55)]

Given the appearance of a crystallographic distortion below T_V that opens an energy gap E_g in the Fe_B-$a_1(\downarrow)$ band, the critical question is whether an analysis of the low-temperature structure and the charge-density distribution within it can provide a guide to the character of the electronic ordering that drives the transition. If the transition is due to a long-range ordering of Fe^{2+} and Fe^{3+} ions, as proposed by Verwey, the cations should remain in the center of symmetry of their interstices. On the other hand, if the transition is due to the formation of a charge-density wave among itinerant Fe_B-$a_1(\downarrow)$ electrons, which interact across shared octahedral-site edges, then some form of metal-metal clustering should be evident. Low-temperature FeS provides a classic example of triangular clustering of high-spin Fe^{2+} ions within a basal plane[57, 58)]. Evidence for localized, majority-spin electrons and itinerant minority-spin electrons in high-spin Fe^{2+} sulfides is strong[59)].

Although early X-ray powder measurements[60)] suggested Fe_3O_4 has rhombohedral symmetry below T_V, magnetic anisotropy measurements on single crystals, first grown by Smiltens[61)], clearly indicated that low-temperature Fe_3O_4 is magnetically orthorhombic. Bickford[30)] demonstrated this with ferrimagnetic resonance and strain-gauge measurements; Calhoun[62)] confirmed it with torque-magnetometer measurements. Moreover, Abrahams and Calhoun[63)] reported an orthorhombic X-ray structure at 78 K with $a_0 = 5.912$, $b_0 = 5.945$, $c_0 = 8.388$ Å, and Hamilton[64)] claimed to confirm the structure with neutron diffraction. At the close of the 1950's, the Verwey model appeared to be established.

The initial difficulty with the model was manifest in attempts to interpret low-temperature Mössbauer data[65)], but the first serious questioning came only in 1968[66)] when electron diffraction from synthetic single crystals at 77 K revealed extra spots that indicated a doubling of the unit cell proposed by Verwey. In 1976, Rado and Ferrari[67] found that Fe_3O_4 exhibits a ferroelectric effect, which requires a noncentrosymmetric space group. Clearly the particular ordering of Fe^{2+} and Fe^{3+} ions proposed by Verwey is not correct.

Four approaches are now taken to Fe_3O_4:

(i) the mobile electrons are considered localized, and the low-temperature data are analyzed in terms of a complex ordering of Fe^{2+} ions arising from different B-site axial fields[68, 69]

(ii) all the 3 d electrons are considered itinerant, but with strong correlations producing high-spin states; the low-temperature structural data is then analyzed in terms of soft phonon modes associated with a charge-density wave[38, 70].

(iii) the mobile electrons are considered localized to Fe_B^{2+}-Fe_B^{3+} pairs above T_V, in which case the low-temperature structure should reflect an ordering of Fe_B-Fe_B dimers[71],

(iv) two mobile electrons are considered to be trapped as bipolarons in molecular-orbital, four-Fe_B-ion clusters of a quadrant of the cubic unit cell, short-range ordering occuring in the interval $T_V < T < 250$ K[72, 73].

Whatever the structural model finally chosen, it must take account of the low-temperature Mössbauer and NMR data. These two types of data are complementary: the NMR signals give a greater accuracy of the values of the hyperfine fields at ^{57}Fe, but the Fe^{2+}-ion signals are weak relative to the Fe^{3+}-ion signals. Mössbauer data, on the other hand, allow quantitative comparison of the signals from Fe^{2+} and Fe^{3+} ions. In addition, a larger isomer shift of the Fe^{2+} relative to the Fe^{3+} lines permits better resolution of the spectra. Both NMR and Mössbauer data exhibit no splitting of the tetrahedral Fe_A^{3+}-ion spectrum below T_V, but a splitting of the Fe_B-atom spectrum into at least four[74–77], but perhaps considerably more, distinguishable iron atoms[78, 79]. The spectra are sensitive to the direction of the applied field and exhibit considerable complexity. Significantly, the orbital momentum remains quenched on all ions, consistent with stabilization of the Fe_B-$a_1(\downarrow)$ orbital by the trigonal component of the crystalline field as shown in Fig. 3. Iida et al.[69] and Mizoguchi[79] have analyzed the low-temperature NMR, Mössbauer and structural data in terms of an ionic model in which Fe^{3+} and Fe^{2+} ions are ordered on the B sites so as to preserve local charge neutrality within a primitive pseudocubic unit cell; account was taken of a highly anisotropic minority-spin charge distribution at the Fe^{2+} ions. However, the Mössbauer isomer shifts are not compatible with simple Fe^{2+} and Fe^{3+} ions; they rather suggest $Fe^{(2+\delta)+}$, $Fe^{(3-\delta)+}$ and $Fe^{(2+\lambda)+}$, $Fe^{(3-\lambda)+}$ pairs with $\delta < \lambda < 0.5$. Moreover, the Iida-Mizoguchi model requires superlattice reflections for h + (1/2), k, l or h, k + (1/2), l, and Izumi et al.[80] point out that these reflections are missing. Although Iida[81] has invoked microscopic antiphase domains in order to salvage the model, some doubt is cast on the uniqueness of this solution; and Cullen[82] has pointed out that the model is probably unstable relative to a partially ordered structure.

The most recent structural data[80] indicate that the low-temperature phase is monoclinic with non-centrosymmetric space group Cc and lattice parameters $\sqrt{2}a_0 \times \sqrt{2}a_0 \times 2a_0$ relative to the cubic-phase a_0. Analysis within the orthorhombic polar space group Pmc 2 identified three soft-mode components that are frozen out below T_V: a dominant mode exhibiting bridging Fe_B–O distances that alternate between shorter and longer for successive bridging oxygens along the b-axis, a Z mode exhibiting Fe_B-Fe_B pairwise motions along [110] and [1$\bar{1}$0] axes of the cubic unit cell, and a polar mode corresponding to clustering of four Fe_B atoms within a quadrant of the unit cell. There is no significant variation in mean Fe_B–O distance, and the maximum moment fluctuation from site to site could be set at less than 0.2 μ_B, so there is no evidence of an ordering of distinguishable Fe_B^{2+} and Fe_B^{3+} ions from the neutron data. It would appear that the electrostatic Coulomb repulsions between the mobile $a_1(\downarrow)$ electrons are accommodated by a long-range ordering of charge transfer between pairs of Fe_B atoms; within a pair, charge transfer is rapid on the time scale of a neutron-diffraction experiment. However, the cation clustering is much less pronounced than occurs in low-temperature FeS where the $a_1(\downarrow)$ electrons are clearly itinerant. In fact, the Mössbauer and NMR data seem to require distinguishable pairs of Fe_B atoms with distinguishable Fe_B sites within each pair on the time scale of 10^{-8} s. The fact that the electron-phonon coupling gives rise to such a complex structure does not necessarily imply that the transition is driven by "nesting" about a Fermi surface, as is commonly assumed for the stabilization of an itinerant-electron charge-density-wave phase. The bandwidth of the Fe_B-$a_1(\downarrow)$ states appears to be small ($w < \Delta_{nc}$), and the electron Hall mobility $u_H < 0.4$ cm²/Vs remains below the lower limit for itinerant electrons. The Verwey transition would therefore seem to be driven by Coulomb repulsions between mobile electrons in a mixed-valence phase that is intermediate between the small-polaron and the itinerant-electron limits.

Finally, Iida et al.[83)] and Rigo et al.[27)] have shown that an anomaly reported at 12 K is due to a small departure from stoichiometry.

2.2.6 *Conclusion*

Magnetite is an inverse spinel $Fe^{3+}[Fe^{3+}Fe^{2+}]O_4$ at room temperature with iron ions containing localized-electron, majority-spin half-shell cores. The splitting $\Delta = \Delta H_{AB} - T\Delta S_{AB}$ between the Fe_A^{3+}/Fe_A^{2+} and Fe_B^{3+}/Fe_B^{2+} couples vanishes at about 1450 °C. The half-shell, high-spin cores on the tetrahedral Fe_A^{3+} ions couple antiparallel to those on the octahedral-site $Fe_B^{3+/2+}$ ions below a Néel ferrimagnetic Curie temperature $T_c \simeq 850$ K, and the dominant Fe_A–O–Fe_B magnetic interaction is well described by superexchange coupling between localized half-shell cores. The mobile Fe_B-$a_1(\downarrow)$ electrons are intermediate in character; they occupy a band that is too narrow for itinerant-electron behavior, and electrostatic Coulomb interactions between the mobile electrons introduce an important electron-phonon coupling below room temperature on the time scale of a neutron-diffraction experiment. This coupling does not produce well defined charge separation into distinguishable Fe_B^{2+} and Fe_B^{3+} ions, as originally proposed by Verwey; nor does it produce recognizable Fe_B^{2+}-Fe_B^{3+} pairs within which the mobile electron is equally shared on a time scale $\tau < 10^{-8}$ s. Rather, at temperatures $T < T_V$, the valence states appear better described by $Fe^{(2+\delta)+}$, $Fe^{(3-\delta)+}$ and $Fe^{(2+\lambda)+}$, $Fe^{(3-\lambda)+}$ pairs with $\delta < \lambda < 0.5$, at least on the time scale $\tau < 10^{-8}$ s of a Mössbauer measurement.

Above room temperature, the mobile 3d electrons are well described by a random mixture of Fe_B^{2+} and Fe_B^{3+} ions with the mobile electrons diffusing from iron to iron, some being thermally excited to Fe_A^{3+} ions, but the motional enthalpy on the B sites is $\Delta H_m < kT$. As the temperature is lowered through T_c, the Seebeck coefficient shows the influence of a change in mobile-electron spin degeneracy, and at room temperature the Seebeck coefficient is enhanced by correlated multielectron jumps that provide a mobile electron access to all its nearest neighbors. The electron-hopping time $\tau_h \simeq \omega_R^{-1} \simeq 10^{-12}$ s

Table 2. Some Properties of Stoichiometric Magnetite

i. *Structure*	Ref.
$T > T_V = 120 \pm 2$ K type $MgAl_2O_4$ (Fd3m) spinel	106
cell parameter: a = 8.3941 Å	106
oxygen parameter: u = 0.3799	80
$T < T_V$ monoclinic (Cc)	48, 49
$dT_V/(T_V dP) = 4 \times 10^{-3}$ kbar^{-1}	
ii. *Electrical properties*	
$\sigma = 250\ \Omega^{-1}\ cm^{-1}$ at 300 K, where it has its maximum value	28
$\lvert u_h \rvert \lesssim 0.1$ cm^2/V.s	42
σ drops from $\simeq 50$ to $\simeq 0.4\ \Omega^{-1}\ cm^{-1}$ at T_V	
$\alpha = -55\ \mu V/°C$ at 300 K; -70 to $-85\ \mu V/°C$ from 820 to 900 K	50
iii. *Magnetic properties*	
Néel collinear ferrimagnet; $J_{AB} \simeq -2.3$ meV, $J_{BB} \simeq 0.68$ meV	32, 34
Curie temperature: $T_c = 850$ K; $dT_c/(T_c dp) = 2.42 \times 10^{-3}$ kbar^{-1}	25, 39
Saturation moment at 0 K: $M_s = 4.1\ \mu_B$/mole	25
Effective spectroscopic splitting factor: g = 2.12	30
iv. *Mössbauer data*	
At 300 K: 2 sextets attributed to A and B sites	99
Isomer shift δ: 0.28 (A) and 0.67 (B) mm/s	
Hyperfine field H: 49.6 (A) and 46.5 (B) Teslas	
Line-width Γ: 0.22 (A) and 0.31 (B) mm/s	
At $T < T_V$: 1 sextet for the A sites (Fe^{3+})	33
2 sextets for Fe^{2+} on B sites	
2 sextets for Fe^{3+} on B sites	

is governed by the optical-mode vibration frequency ω_R. Since the jump distance is only to a nearest neighbor, but with $\Delta H_m < kT$, the conductivity is nearly temperature-independent except for spin-disorder scattering near T_c that may represent the formation of magnetic polarons.

In the temperature interval $T_V < T < 250$ K, strong electron-phonon coupling reduces the electrostatic Coulomb interactions between mobile electrons via a short-range, dynamic ordering; the nature of the optical modes that become soft suggests that the mobile electrons may become ordered into Fe_B^{3+}-Fe_B^{2+} pairs, but with an important polar component on a short time scale ($<10^{-12}$ s). Where two such pairs coincide on the same four-Fe_B-ion tetrahedron, two mobile electrons would be trapped in what might be described as a parallel-spin bipolaron. The symmetries of the modes that soften emphasize the intermediate character of the mobile electrons. The semiconductor-semiconductor transition at T_V is clearly different from the semiconductor-metal transition occuring in materials like VO_2 or Ti_4O_7.

Given this general picture for Fe_3O_4, it is instructive to see how it is modified on going away from stoichiometry on substitution of foreign ions.

2.3 *Iron-Deficient Magnetite*

The system $Fe_{3-x}O_4$ can be prepared at low temperatures as a single spinel phase over the entire compositional range $0 \leqslant x \leqslant 0.33$[22]. The end number, γ-Fe_2O_3, is known as maghemite. At higher temperatures, the system disproportionates into Fe_3O_4 and α-Fe_2O_3; the end member γ-Fe_2O_3 transforms to α-Fe_2O_3, which has the corundum structure, at about 400 °C[84].

Since the spinel phase must be prepared at low temperatures (by hydrothermal synthesis or by careful oxidation of magnetite at a temperature $T < 300$ °C, for example) it has been widely suspected that some incorporation of hydrogen is needed to stabilize it. However, Schrader and Büttner[22] have shown that pure γ-Fe_2O_3 does exist, and Coey et al.[85] have been able to prepare $Fe_{3-x}O_4$ in the compositional range $0 < x < 0.08$ by quenching non-stoichiometric magnetite prepared at 1450 °C. There is no evidence that hydrogen is needed to stabilize the system.

On the other hand, a non-equilibrium synthesis does tend to produce samples that may be chemically inhomogeneous or have a non-reproducible distribution of cation vacancies. For example, it has been established[86–88] that oxidation occurs by diffusion of iron out into oxygen layers adsorbing on the surface, so it has been suggested[89] and proven[90] that preparation of γ-Fe_2O_3 by the low-temperature oxidation of fine particles of Fe_3O_4 may yield an Fe_3O_4 core with a γ-Fe_2O_3 skin. Moreover, samples prepared in different laboratories tend to have different distributions of the cation vacancies. Although the equilibrium distribution in γ-Fe_2O_3 has all the vacancies ordered on B sites to give a superstructure of the cubic spinel[91], nevertheless under some preparative conditions the vacancies may be partially distributed on the tetrahedral sites[71]. For example, Henry and Boehm[92] concluded, from magnetization data and the assumption of a Néel ferrimagnetic order, that the vacancies were exclusively on B sites in their samples whereas the neutron-diffraction data of Ferguson and Hass[93] showed 10-percent vacancies on A sites in their sample and Takei and Chiba[94], from X-ray data, found 20-percent A-site vacancies in a single-crystal sample of γ-Fe_2O_3.

The introduction of Fe_B-atom vacancies modifies the energy diagram of Fig. 3. The preservation of local charge neutrality requires that each Fe_B-atom vacancy have five of its six nearest neighbors as Fe_B^{3+} ions, which means that the $a_1(\downarrow)$ states of five Fe_B ions neighboring a vacancy, $\square$, are raised an energy E_A above the Fermi energy. They become acceptor states inaccessible to the mobile electrons at low temperatures. It is possible to express this situation with the structural formula

$$Fe_{3-x}O_4 = Fe^{3+}[(\square Fe_5^{3+})_x Fe_{1-3x}^{2+} Fe_{1-3x}^{3+}]O_4 \tag{17}$$

As pointed out by Coey et al.[85], such a description can account well for two observations:

(i) T_V decreases with x and

(ii) the Mössbauer room-temperature spectra show three lines with isomer shifts corresponding to Fe_A^{3+}, Fe_B^{3+} and $Fe_B^{2.5+}$.

The significant point is that the bulk Fe_B matrix seen by the mobile $a_1(\downarrow)$ electrons retains a valence ratio $Fe_B^{2+}/Fe_B^{3+} = 1$. As the volume of this matrix decreases, so does T_V, but the basic electrostatic Coulomb repulsions within the matrix remain the same, so the electron-phonon coupling may retain its essential character. The initial analysis of the room-temperature Mössbauer spectra – for x = 0.06, 0.10, 0.18, 0.26 and 0.33 – in terms of three six-line patterns[95] was assumed to be proof of the formation of Fe_B^{3+}-Fe_B^{2+} pairs as originally proposed by Daniels and Rosencwaig[71], but such a model is not a unique interpretation of the data. For x = 0.18, for example, the isomer shifts were $\delta = 0.27$, 0.39, and 0.65 mm/s wrt iron; and the internal effective fields were $H_{eff} = 498.6$, 501.9 and 420.3 kOe, respectively. From Eq. (1), these would correspond to Fe_A^{3+}, Fe_B^{3+} and $Fe_B^{2.5+}$. These results were essentially confirmed by Coey et al.[85], who used a 47 kOe field applied parallel to the γ-ray propagation direction to improve the resolution. Their peak intensities and isomer shifts were in good agreement with formula (17). They differed from the earlier study only in a measured $H_{eff} = 463$ kOe for the $Fe_B^{2.5+}$ ions.

Gillot[96] has measured the A.C. conductivity at 1 kHz of samples with x = 0.015, 0.025, and 0.04 from 77 to 273 K. A continuous decrease in T_V with increasing x indicated a solid solution of vacancies in his samples. The semiconductor-semiconductor change across T_V also decreased with increasing x primarily due to an increase in the activation energy of the high-temperature phase below 273 K. Gillot interprets this evolution in the activation energy above T_V as a manifestation of stronger electron-phonon coupling, which is reasonable in view of the perturbation of the Fe_B-$a_1(\downarrow)$ band by the vacancies. Moreover, the ($\square Fe_B^{3+}$) cluster would not cooperate with a long-range soft mode, so short-range ordering would be strengthened. Gillot opted for Fe^{2+}-Fe^{3+} pairing as the most probable short-range unit.

2.4 *Substituted Magnetite*

2.4.1 *Fluorine Substitution: $Fe_3O_{4-x}F_x$*

Portier et al.[97] were the first to introduce F^- for O^{2-} ions in magnetite. From magnetic data and the assumption of Néel ferrimagnetism, they were able to establish that the low-temperature chemical formula is $Fe^{3+}[Fe_{1-x}^{3+}Fe_{1+x}^{2+}]O_{4-x}F_x$. Since the F^- ion is more ionic, the mean Fe_A-X-Fe_B superexchange interaction decreases and so T_c decreases with increasing x in conformity to the observation of a $dT_c/dP > 0$ in Fe_3O_4. Since the F^- ion is about the same size as the O^{2-} ion, the addition of Fe_B-$a_1(\downarrow)$ electrons causes the cell parameter a_0 to increase with x; but the room-temperature lattice parameter does not increase linearly with x, which suggests that at this temperature an increasing fraction of the mobile electrons are thermally excited to the Fe_A atoms. This latter observation indicates a decrease with x in the energy Δ of Fig. 3, which should be expected as the energy Δ in Fe_3O_4 arises from the stronger Fe_A–O covalent mixing relative to the Fe_B–O covalent mixing; electrostatic considerations above favor a normal spinel with Fe_A^{2+} ions. In fact, each F^- ion must trap an Fe_B^{2+} ion at a donor energy E_D below E_F as well as create empty localized states associated with nearest-neighbor Fe_A and Fe_B ions. The result would be a "band tailing" of localized states below a mobility edge for the narrow Fe_B-$a_1(\downarrow)$ band and a similar tailing of localized Fe_A-$e(\downarrow)$ states below the empty Fe_A-$e(\downarrow)$ level. The density of these "tail" states would increase with x.

The fact that the temperature domain of positive Seebeck coefficients below T_V, see Fig. 6, is wiped out by x = 0.01 shows that a donor state associated with an F^- ion lies above the top of the Fe_B-$a_1(\downarrow)$ valence band in the low-temperature phase; it charge compensates the holes in that band. This, in turn, means that the clusters of an F^- ion

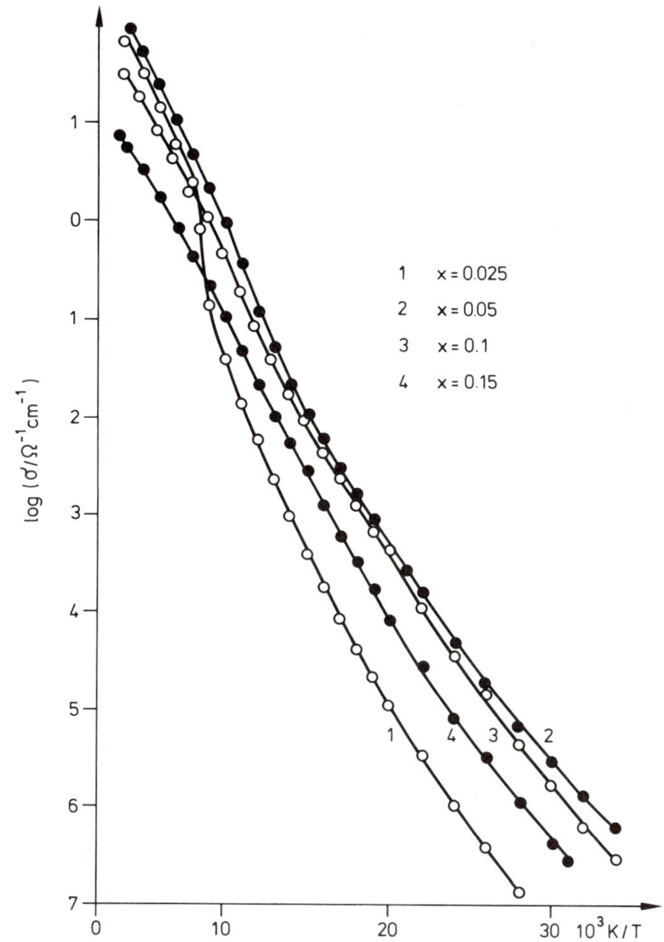

Fig. 7. Conductivity of $Fe_3O_{4-x}F_x$ versus reciprocal temperature (20 < T < 300 K), after[55]

and four nearest-neighbor iron (Fe_A + 3 Fe_B) do not participate in the Verwey transition, so T_V and the conductivity change across it decrease with increasing x, apparently vanishing for x > 0.025, see Fig. 7[55].

On the other hand, above T_V the Seebeck coefficient remains negative despite the increasing density of Fe_B^{2+} ions, see Fig. 5. This can be attributed to two factors:
(i) a trapping of each added $a_1(\downarrow)$ electron to its parent F^- ion and
(ii) the greater accessibility of Fe_A atoms.

The "tail" states associated with F^--ion substitutions also introduce the possibility of "variable-range hopping" and hence a $\ln\sigma$ versus $T^{-1/4}$ dependence. Indeed, Graener et al.[55] have observed a $T^{-1/4}$ law at lowest temperatures in their F^--doped samples.

Further evidence for trapped Fe_B^{2+} ions comes from Mössbauer data taken at 300 K and 77 K[98–100], which show identifyable Fe_B^{2+} ions; the intensity of the weak peak (relative to the averaged Fe^{2+}, Fe^{3+} signal) increases with x. In addition, the B-site signal characteristic of pure magnetite is broadened by the F^--ion substitutions, which is probably due to an enhancement of the quadrupole splittings rather than a decrease in the hopping time τ_h.

As in the system $Fe_{3-x}O_4$, the washing out of the semiconductor-semiconductor transition at T_V, but the retention of strong electron-phonon coupling to above room temperature, shows that the long-range order below T_V in magnetite is only a freezing in of a dynamic coupling established at much higher temperatures; this coupling is presumably, driven by electrostatic Coulomb interactions between the mobile electrons. Such couplings are not associated with itinerant 3 d electrons in metallic oxides.

2.4.2 Substitution on A Sites

Whether a solute cation substitutes for A-site or B-site iron depends upon the relative site-preference energies of the ions. Moreover, a solute atom may lower the energy Δ of Fig. 3 to where significant concentrations of Fe_A^{2+} ions are stabilized in the presence of Fe_B^{3+} ions at room temperature. Temperature-dependent cation distributions can be expected where the relative site-preference energies differ by a $\Delta\varepsilon < kT$.

Various attempts have been made to calculate the site-preference energies for spinels, and a tabulation such as that given by Paul and Basu[101)] can be a useful guide if not taken too literally. As a general rule, electrostatic energies favor stabilization of ions with smaller charge on the A site. However, smaller cations with important (sp^3) covalent mixing dominating any d-orbital covalent mixing also tend to be stabilized in A-sites, and covalent mixing is more important on ions of higher formal valence. The Fe^{3+} and Mn^{2+} ions, which have half-filled, high-spin 3 d shells, have (sp^3) covalent mixing dominant. Of the ions with partially filled d shells, crystalline fields give a strong B-site preference to ions with d^3, low-spin d^5 or d^6, or high-spin d^8 configurations. Solute cations that can compete with Fe^{3+} ions for the tetrahedral sites include: (Cu^+), (Mg^{2+}), Zn^{2+}, Cd^{2+}, (Mn^{2+}), (Co^{2+}), (Ga^{3+}), (In^{3+}), Si^{4+}, (Ge^{4+}), and the Si^{4+} ion tends to stabilize other phases. The ions in parentheses do not substitute exclusively on A-sites in the presence of Fe^{3+} ions.

Divalent Solutes. The cations Zn^{2+} and Cd^{2+} have a strong enough tetrahedral-site preference to displace only the A-site iron if the samples, prepared at high temperature, are cooled slowly enough for low-temperature equilibrium to be established. The most extensive work has been done on the system Fe_3O_4-$ZnFe_2O_4$. Conclusions drawn from the data on this system appear to apply also to the system $Cd_xFe_{1-x}[Fe_2]O_4$.

Although Popov and Ilinova[102)] have reported deviations from Végard's law in their preparations of the system Fe_3O_4 in $ZnFe_2O_4$, other workers[103–105)] have obtained a linear variation of lattice parameter with Zn concentration x, from $a_0 = 8.3941$ Å with $u = 0.3799$ for Fe_3O_4[106)] to $a_0 = 8.4411$ Å[25)] with $u = 0.380$[107)] for $Zn[Fe_2]O_4$. The formulation $Zn_x^{2+}Fe_{1-x}^{3+}[Fe_{1+x}^{3+}Fe_{1-x}^{2+}]O_4$ appears to be nearly attainable with careful preparations.

Substitution of a nonmagnetic divalent cation onto the A sites has three immediate consequences:

(i) The Curie temperature T_c decreases as the density of Fe_A–O–Fe_B interactions is reduced

(ii) The Fe_B^{2+}-ion concentration is reduced
(iii) The mobile-electron potential is perturbed

As T_c decreases, the Fe_B–Fe_B interactions become more competitive with the Fe_A–O–Fe_B interactions. Moreover, Zn-rich regions become depleted of mobile Fe_B^{2+} ions; what Fe_B^{2+} ions remain in these regions tend to be trapped as small polarons because the density of mobile carriers is too low to inhibit polaron formation and the perturbation of the electronic potential tends to make neighboring sites energetically inaccessible at the lower temperatures. The ferromagnetic double-exchange component to the Fe_B–Fe_B interactions requires a charge transfer (without change of spin) that is fast relative to the time $\tau_r \sim 10^{-6}$ s for the total-electron spin to relax to a different orientation. A small-polaron formation increases $\tau_h \sim \omega_R^{-1} \exp(\Delta H_m/kT)$ by an exponential factor for $\Delta H_m > kT$, which tends to make $\tau_h > \tau_r$. Therefore, the Fe_B–Fe_B interactions change from ferromagnetic to antiferromagnetic. Where these interactions compete with the Fe_A–O–Fe_B interactions, which are also weaker in the zinc-rich regions, a "frustration" of the magnetic couplings can occur. This is the condition for forming a "spin glass".

From these simple considerations, it follows that for low values of x, the spontaneous magnetization should vary with x according to the Néel theory of ferrimagnetism, see Eq. (3):

$$\begin{aligned} \bar{\mu}_{mol} &= (1+x)\mu_{Fe(III)} + (1-x)\mu_{Fe(II)} - (1-x)\mu_{Fe(III)} \\ &= (1-x)\mu_{Fe_3O_4} + 10\,x\,\mu_B \end{aligned} \qquad (18)$$

where μ_B is the Bohr magneton and $\mu_{Fe(III)} = 5\,\mu_B$ has a spin-only value. It has been known since the early work of Guillaud[108] and Gorter[109] that the saturation magnetization of the system $Zn_x^{2+}Fe_{1-x}^{3+}[Fe_{1+x}^{3+}Fe_{1-x}^{2+}]O_4$ is in essential agreement with the Néel theory for $x < 0.5$, but with a modest reduction from the Néel theory, see Fig. 8. The origin of this deviation is not established. Although the Mössbauer data are compatible with some canting of the B-site moments from collinearity[104], this explanation is improbable in view of the significant ferromagnetic Fe_B–Fe_B coupling found in Fe_3O_4. It is more probably due to some small deviation from the ideal Zn^{2+}-ion distribution that is not readily detected by diffraction techniques.

A sharp drop in the spontaneous magnetization for larger values of x indicates a critical concentration in the range $0.5 < x_c < 0.6$ where small-polaron formation in Zn-rich regions begins to produce spin-coupling "frustrations", and a spin-glass behavior can be anticipated for compositions in the range $0.5 < x < 1.0$. Mössbauer studies[110–113] have established that, at room temperature, $\tau_h < 10^{-7}$ s is found in the compositional range $x < 0.6$ and that paramagnetic and magnetically ordered phases coexist in the compositional range $0.6 < x < 0.8$; only paramagnetic iron exists at room temperature for $x > 0.8$, but at 4.2 K the iron are ordered and the spectrum is complex[113].

Ok and Evans[112] have further shown that the room-temperature isomer shift changes linearly with x over the range $0 \leq x \leq 0.3$ as predicted from Eq. (1) for an access of the mobile electrons to all the B-site iron atoms. This finding argues strongly against the formation of Fe_B^{2+}–Fe_B^{3+} pairs at room temperature, in agreement with our deductions for Fe_3O_4.

In contrast to $Fe_{3-x}O_4$ and $Fe_3O_{4-x}F_x$, the Verwey transition disappears even for small concentrations of zinc in the system $Zn_xFe_{3-x}O_4$[114]. This observation indicates that

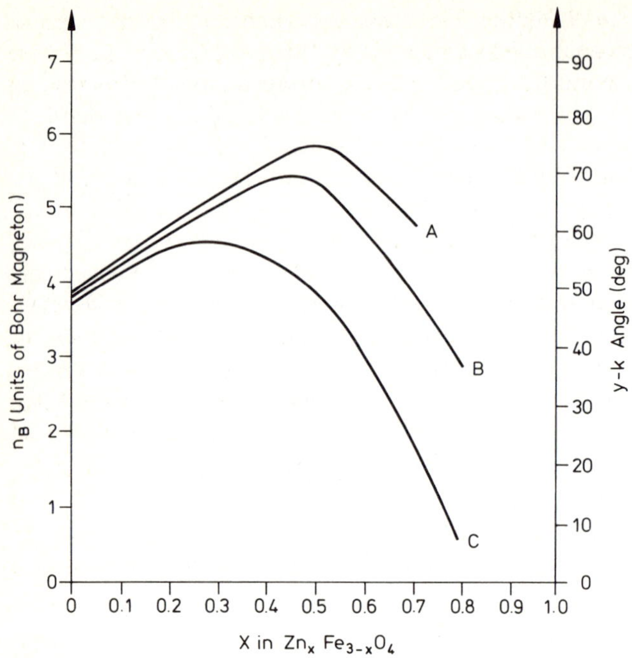

Fig. 8. Comparison of Néel theory and measured Bohr magneton number $n_B = \bar{\mu}_{mole}/\mu_B$ at T = 5 K (*A*), 77 K (*B*), and 303 K (*C*) for $Fe_{3-x}Zn_xO_4$, after[104)]

the Verwey transition is associated with a valence ratio $Fe_B^{2+}/Fe_B^{3+} \simeq 1.0$ on the B sites, which favors a long-range ordering of the mobile Fe_B-$a_1(\downarrow)$ electrons as a result of electrostatic Coulomb interactions between the mobile electrons and/or electron-phonon interactions. In the former two systems, this ratio is retained in the bulk matrix, as was discussed above, whereas in $Zn_xFe_{3-x}O_4$ it is not, according to the Mössbauer data of Ok and Evans[112)].

The fact that the Fe_B^{2+}/Fe_B^{3+} ratio changes with x in the system $Zn_xFe_{3-x}O_4$ permits exploration of the conclusion made above that Eq. (16) is more appropriate than Eq. (15) for the room-temperature Seebeck data for Fe_3O_4. Srinavasan and Srivastava[45)] have emphasized that in Zn-substituted magnetite, the Seebeck coefficient is negative and nearly temperature-independent in their samples over the temperature range $200 < T < 500$ K; moreover, its magnitude is given by Eq. (16) – not (15) – so that

$$\alpha = -(k/e)\ln|2/(1-x)| \tag{19}$$

in the compositional range $0 < x < 0.4$. They argue that this relationship demonstrates a phonon-induced tunneling of electrons from one B-site to another, which corresponds to a diffusional motion with $\Delta H_m < kT$. In this model, the mobile electrons have access to every Fe_B nearest neighbor, which must imply the possibility of multielectron jumps correlated by the electron-phonon interactions. Whether there is a smooth transition to the compositional dependence of Eq. (16) at temperatures $T > T_c$, as suggested by the data of Figs. 4 and 5 for Fe_3O_4, needs clarification.

Trivalent Solutes. It has proven difficult to find a trivalent cation that substitutes exclusively for Fe_A^{3+} ions over a complete solid-solution range. However, Dehe et al.[115] have compared Mössbauer, neutron-diffraction, and saturation magnetization data to arrive at the following formulations:

$$\text{For } x < 0.3 \qquad Fe_{1-x}^{3+}Ga_x^{3+}[Fe^{2+}Fe^{3+}]O_4 \tag{20}$$

$$\text{For } 0.3 < x < 2.0 \qquad Fe_{1-x+\alpha}^{3+}Ga_{x-\alpha}^{3+}\,[Fe^{2+}Fe_{1-\alpha}^{3+}Ga_\alpha^{3+}]O_4$$

with $\alpha \simeq 0.6\,(x - 0.3)$, in rough agreement with older determinations[116, 117]. The Ga^{3+} ions appear to be randomly distributed over energetically equivalent A and B sites for $x > 0.3$. Although the samples were quenched from 1050 °C, subsequent heat treatments (unspecified) had little apparent influence on the Ga^{3+}-ion distribution in the range $0 < x < 0.8$.

Room-temperature Mössbauer data revealed that fast electron transfer ($\tau_h < 10^{-8}$ s) is retained throughout the compositional range $0 < x < 0.8$. Samples with $x = 0.4$ and 0.8 exhibited room-temperature neutron-diffraction data consistent with collinear Fe_A^{3+}-ion spins below T_c, but with a significantly reduced moment (ca. 1 μ_B vis à vis a saturated spin-only value) on the B sites. However, at 4.2 K, the spontaneous magnetization could be interpreted with a Néel ferrimagnetic model:

$$\bar{\mu}_{mol} = (4 - 10\,\alpha + 5\,x)\mu_B \tag{21}$$

with a spin-only value of 4 μ_B for the Fe_B^{2+} ions. The relative slowness of the Fe_B ions to achieve a saturation magnetization with decreasing temperature simply reflects the fact that every Fe_A^{3+} ion sees a nearly full complement of Fe_B-ion nearest neighbors to couple to by Fe_A–O–Fe_B superexchange whereas each Fe_B ion sees a significantly reduced number of Fe_A ions.

Substitution of In^{3+} for Fe^{3+} in Fe_3O_4 gives an even smaller compositional range ($x < 0.15$) for exclusive A-site substitutions[118, 119], but preliminary investigations indicate an influence on the properties of Fe_3O_4 similar to that of Ga substitution.

2.4.3 Substitution on B Sites

Substitution exclusively on the B sites can be achieved with ions having formal valence ranging from I to VI. It is instructive to consider representative samples.

Monovalent Solute. The system $Fe^{3+}[Li_x^+\,Fe_{1-2x}^{2+}Fe_{1+x}^{3+}]O_4$ exists over the complete compositional range $0 \leqslant x \leqslant 0.5$. The Li^+ ions are ordered on the B sites of the end number $x = 0.5$[120]; several ordered arrangements are possible depending on the thermal treatment[121]. The stronger (sp^3) covalence at an Fe_A^{3+} ion displaces the lower-valent Li^+ ion to an octahedral site.

Lithium ferrite itself ($x = 0.5$) has a high Curie temperature and can be fabricated so as to give a square hysteresis loop satisfactory for digital-computer "memory" cores. In this application, the dielectric losses connected with the presence of mobile charge carriers can cause a dramatic loss in core quality. The mobile carriers may be introduced by impurities or by loss of Li_2O during sintering. A practical solution to the latter problem is to substitute a small fraction of the iron with manganese, which traps any mobile Fe_B-$a_1(\downarrow)$ electrons via the reaction

$$Fe_B^{2+} + Mn^{3+} \rightarrow Fe_B^{3+} + Mn^{2+} \quad (22)$$

Any mixed Mn^{2+}, Mn^{3+} valency due to overcompensation for Li_2O loss is inconsequential so long as the manganese concentration is kept too low for any significant electron transfer between solute atoms. Belov et al.[122] have used Mössbauer measurements to demonstrate this function for the manganese. The role of manganese is similar in more complex systems where Li_2O loss would otherwise give excessive dielectric loss[123, 124].

Divalent Solute. The $3d^8$ configuration at a Ni^{2+} ion gives it a strong octahedral-site preference, so the system $Ni_xFe_{3-x}O_4$ $(0 \leq x \leq 1)$ has, to a good approximation, the cation distribution $Fe^{3+}[Ni_x^{2+}Fe_{1-x}^{2+}Fe^{3+}]O_4$.

The electrical and magnetic properties of this system have been extensively studied, and the results are well summarized in available reviews[91, 125]. It is a Néel ferrimagnet for all x, and the mobile-electron hopping time τ_h decreases with increasing x; small polarons with a $\Delta H_m > kT$ and a $\tau_h > 10^{-8}$ s are found for $x > 0.2$. The activation energy varies with x essentially through the concentration of Fe_B^{2+} ions[126], as it does generally for the substituted magnetites[127]. As the concentration of Fe_B^{2+} ions decreases, so does the inhibition of polaron formation imposed by polaron-polaron interactions.

The Seebeck coefficient α becomes temperature-independent only above a temperature T_t, where $T_t \simeq 300$ K for $x < 0.1$[128, 129]. In magnetite, T_t can be identified with the onset of strong electron-phonon coupling. The temperature-independent α shows a continuous evolution from the value for $\beta = 1$ described by Eq. (16) at $x = 0.1$ to that by Eq. (15) for $x > 0.8$. Although small-polaron formation is observed for $x > 0.2$, regions apparently persist where multielectron jumps can occur.

At lower temperatures $(T < T_t)$, α exhibits a temperature dependence characteristic of a small activation energy ($\simeq 0.03$ eV) for excitation of charge carriers from stationary trap sites[130]. It is reasonable to suspect that small polarons tend to be trapped at impurity centers at low temperature.

Direct evidence for an inhomogeneous distribution of small polarons and electrons in fast-charge-transfer states is found in the room-temperature Mössbauer spectra of Linnett and Rahman[130]; a discontinuity in the resistivity appears only for $x > 0.5$ whereas the more complex Mössbauer spectrum is already present at $x > 0.2$. In summary, the data show for part of the iron an increase in τ_h with x, a crossing of the 10^{-8} s Mössbauer timescale occuring in the range $0.2 < x < 0.4$; but the region of fast-electron transfer exceeds the percolation limit for $x < 0.5$. Interpretation of the Mössbauer spectrum in the interval $0.4 < x < 0.8$ has been based on Eq. (1) for the Fe_B isomer shifts within regions or clusters sustaining a $\tau_h < 10^{-8}$ s. These regions correspond to $Fe_B^{2.5+}$ only at $x \lesssim 0.4$; an Fe_B^{n+} having $n > 2.5$ is observed for larger x.

Trivalent Solute

(a) *Cr^{3+}:* the strong crystal-field stabilization for the d^3 configuration of a Cr^{3+} ion places this cation unambiguously on the B sites. However, the $2p^6$ configuration at an O^{2-} ion is sufficiently polarized toward the Cr^{3+} ion – via the σ-bond covalent component with the empty Cr^{3+}-e orbitals, which is what produces the crystal-field stabilization – that the (sp^3) covalent mixing at neighboring Fe_A^{3+} ions is reduced. This latter reduction causes the energy Δ of Fig. 3 to be lowered at the Fe_A iron nearest-neighbor to a Cr_B^{3+} ion, and as the number of nearest-neighbor Cr_B^{3+} ions increases, Fe_A^{2+} ions become stabilized. Thus for larger x the local chemical inhomogeneities tend to produce, in this case, Fe_A^{2+} ions in Cr_B^{3+}-rich regions and Fe_A^{3+} ions in Cr_B^{3+}-poor regions.

The first indication that this might be the case comes from extreme deviations from Végard's law, the a_0 vs x plot indicating four distinct linear regions of widely different slope[131,132], see Fig. 9 (a):

I $= 0 \leqslant x < 0.3$
II $= 0.3 < x < 0.7$
III $= 0.7 < x < 1.3$ and
IV $= 1.3 < x \leqslant 2.0$

These data suggest a general formulation

$$Fe_{1-y}^{3+}Fe_y^{2+}[Fe_{1-y}^{2+}Fe_{1-x+y}^{3+}Cr_x^{3+}]O_4 \tag{23}$$

Mössbauer data at 4.8 K, taken with and without an external field of 48 kOe parallel to the direction of the γ rays, has established a statistical distribution of the Cr_B^{3+} ions on the B sites and, at 4.8 K, a valency distribution $y = 0$ over regions I and II[133]. Two Fe_A and four F_B hyperfine fields could be resolved; the four Fe_B spectra represent Fe_B ions having 0, 1, 2 and 3 nearest-neighbor Cr_B^{3+} ions. Significantly, an increase in the Fe_B^{2+} character, as determined from the isomer shift and Eq. (1), with increasing number of ${Cr_B}^{3+}$ neighbors was also evident, as required for local charge neutrality. This observation argues against Fe_B^{2+}–Fe_B^{3+} pair formation even at 4.8 K in this system, in contrast to conclusions reached earlier by Robbins et al.[134]. Room-temperature Mössbauer data[133] for $0 < x < 0.5$ also demonstrated a statistical distribution of Cr_B^{3+} ions, but indicated a $y = 0$ only in Region I at 300 K. It is thus clear that Δ of Fig. 3 is reduced for Fe_A ions having Cr_B^{3+}

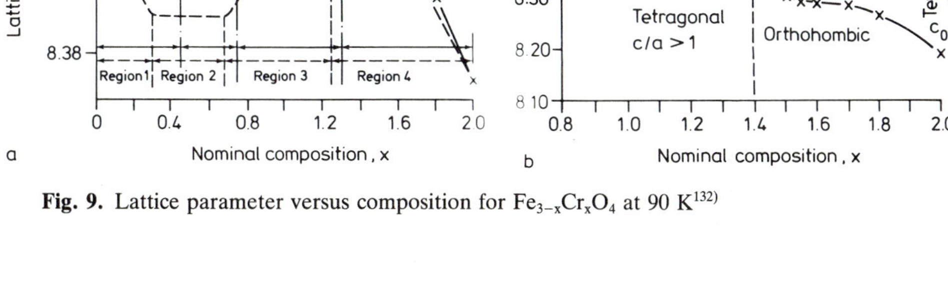

Fig. 9. Lattice parameter versus composition for $Fe_{3-x}Cr_xO_4$ at 90 K[132]

near neighbors and that stabilization of Fe_A^{2+} ions at lowest temperatures (y > 0) is established throughout Region III. It follows that Δ changes sign where an Fe_A ion has half or more of its nearest neighbors Cr_B^{3+} ions.

Francombe[132] performed powder X-ray diffraction measurements at 90 K and observed a cooperative tetragonal (c/a > 1) distortion in the range $0.9 < x < 1.3$, a tetragonal (c/a < 1) distortion in $Fe[Cr_2]O_4$, and an orthorhombic phase in the range $1.3 < x < 1.95$, see Fig. 9(b). The distortion in $Fe[Cr_2]O_4$ can be attributed to a cooperative Jahn-Teller distortion associated with Fe_A^{2+} ions, which have a ^5E_g ground state in the cubic phase[135]. The tetragonal (c/a > 1) distortion can be assigned to a cooperative Jahn-Teller distortion that preserves the spin-orbit coupling at Fe_B^{2+} ions located in sites where the sign of the local trigonal component of the B-site crystalline field has been reversed and fast electron transfer has given way to small-polaron formation[40, 136]. In this latter case, the crystallographic transition must occur below a magnetic ordering temperature with collinear B-site spins; magnetic order provides the cooperative alignment of spins, and spin-orbit coupling provides the orbital ordering responsible for the distortion. In $Fe[Cr_2]O_4$, where the crystallographic transition lies above the magnetic-ordering temperature, a critical Fe_A^{2+}-ion concentration must be reached for a cooperative, static distortion to occur. This critical concentration should be close to one Fe_A^{2+} ion per mole. The orthorhombic distortion is a superposition of the two tetragonal distortions, so we may conclude that in Region IV ($1.3 < x < 2.0$), a $y = 1 - \delta$, where δ is a small fraction that decreases with increasing x, must be used in Eq. (23). For the tetragonal (c/a > 1) distortion due to Fe_B^{2+} ions, a critical concentration is not required; however, the Fe_B^{2+} ion must have a localized $^5T_{2g}$ configuration. Therefore, we may conclude that small-polaron formation with a $\tau_h > 10^{-8}$ s must be found for x > 0.9.

Magnetic data for the entire compositional range $0 < x < 2.0$ are given in Fig. 10[134]. In compositional Regions I and II, the spontaneous magnetization is only slightly reduced from the Néel ferrimagnetic model for spin-only atomic moments; in Region III, a Néel ferrimagnetic model begins to break down with increasing x. In this region, slowing down of the electron-transfer hopping frequency eliminates the ferromagnetic

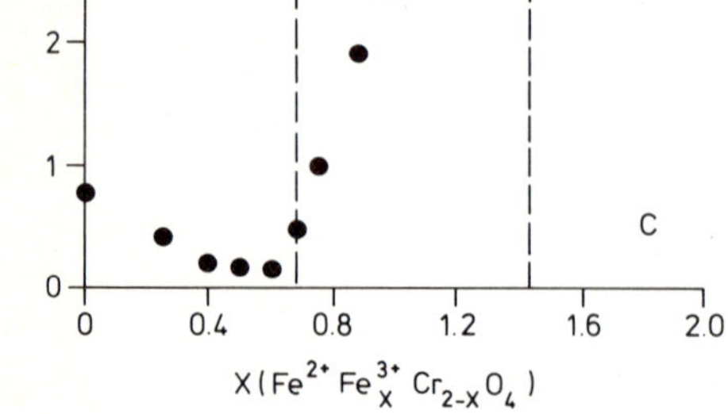

Fig. 10. Comparison of Néel theory and measured Bohr magneton number $n_B = \bar{\mu}_{mole}/\mu_B$, extrapolated to T = 0 K, for $Fe_{3-x}Cr_xO_4$, after[134]

double-exchange component to the Fe_B–Fe_B interactions, and all the B–B interactions become antiferromagnetic. Competition between A–O–B and B–B interactions introduces "frustrations" and a spin glass. In the tetragonal ($c/a < 1$) phase of $Fe[Cr_2]O_4$, a Yafet-Kittel triangular-spin configuration[137] is stable[15, 138] and has been established with neutron diffraction[139].

(b) *Al^{3+}:* The Al^{3+} ion substitutes almost exclusively on B sites in the system $Al_xFe_{3-x}O_4$, $0 \leqslant x \leqslant 2$. For $x = 2$, the free energy to move an Al^{3+} ion from a B site to an A site contains a $\Delta H° = 0.53$ eV and a $\Delta S°/R = 0.63$[140]; and with low-temperature (900 °C) synthesis, the A sites are only 2.5-percent occupied by Al^{3+} ions[141]. Mössbauer data from 77 to 647 K[142] confirm the retention of a $\tau_h < 10^{-8}$ s for $x < 0.9$ and a random distribution of Al^{3+} ions on the B-sites. Since the B-site isomer shift remains nearly constant at a value corresponding to $Fe_B^{2.5+}$, according to Eq. (1), the cation distribution has been described by the formulation:

$$Al_\alpha^{3+}Fe_{1-\alpha-y}^{3+}Fe_y^{2+}[Al_{x-\alpha}^{3+}Fe_{1-y}^{2+}Fe_{1-x+\alpha+y}^{3+}]O_4 \qquad (24)$$

with

$$y = (x/2) - \alpha \quad \text{and} \quad \alpha \simeq a(T)x \qquad (25)$$

where $a(1000\,°C) \simeq 0.05$ at $x = 2$. A nearly linear shift of Fe^{2+} ions from B to A sites with increasing x is in surprising contrast with the system $Fe_{3-x}Cr_xO_4$; it suggests a stabilization of an octahedral-site valence ratio $Fe_B^{2+}/Fe_B^{3+} = 1$ as a result of electron-phonon interactions. However, high-temperature Seebeck data, albeit on oxygen-deficient samples prepared under low oxygen pressure at high (1280–1500 °C) temperature, give a change in sign at $x \simeq 1/3$ that can only be satisfactory interpreted in terms of diffusional hopping of mobile electrons between all the iron atoms, both Fe_A and Fe_B[143].

The paramagnetic susceptibility of pure $FeAl_2O_4$ ($x = 2$) gives a Weiss constant $\theta = -144$ K and a molar Curie constant (emu) $C_m = 3.80$ K[144], somewhat large for high-spin Fe_A^{2+} ions with a spin-only moment; however, there is no apparent antiferromagnetic ordering of the Fe_A^{2+}-ion spins down to 9.5 K, where a peak in the susceptibility has been associated with a local Jahn-Teller distortion about the Fe_A^{2+} ions[145].

(c) *V^{3+}:* A large 3d component in the covalent mixing with V^{3+} ions stabilizes vanadium unambiguously on the B sites in the system $Fe[V_xFe_{2-x}]O_4$, $0 \leqslant x \leqslant 2$. However, there has been some confusion about the valence state of vanadium as well as of the valency distribution among the iron atoms. For example, the composition Fe_2VO_4 has been variously postulated to be

$Fe^{3+}[Fe^{2+}V^{3+}]O_4$[146–148]

$Fe^{2+}[Fe^{2+}V^{4+}]O_4$[149, 150]

$Fe_{0.54}^{2+}Fe_{0.46}^{3+}[Fe_{0.46}^{2+}Fe_{0.54}^{3+}V^{3+}]O_4$[151]

$Fe_{0.4}^{2+}Fe_{0.6}^{3+}[Fe_{0.5}^{2+}Fe_{0.5}^{3+}V_{0.1}^{2+}V_{0.9}^{3+}]O_4$[152]

Mössbauer data[152] are able to rule out the distribution $Fe^{2+}[Fe^{2+}V^{4+}]O_4$, which is consistent with the observation[9] of Fe^{3+} and V^{3+} ions coexisting on equivalent octahedral sites in $FeVO_3$ and with placement of the V^{4+}/V^{3+} redox couple significantly below the Fe^{3+}/Fe^{2+} couple for octahedral solutes in TiO_2[3]. Since the introduction of trivalent solutes has been seen above to introduce a transfer of Fe^{2+} ions from B to A sites and the Mössbauer data are consistent with a $\tau_h < 10^{-8}$ s and a mean valence $Fe_B^{2.5+}$ according to Eq. (1), the last two distributions cited above become the most plausible alternatives. The possibility of a mixed V^{3+}/V^{2+} valence coinciding with a mixed Fe^{3+}/Fe^{2+} valence must be seriously entertained as the energy difference between the two couples appears to be small[3]. Overlap of the redox potentials is possible in solids where each redox couple is represented by a finite bandwidth.

With reference to the energy diagram of Fig. 3, the complex valence distribution shown in the last alternative above would be represented by an overlap of partially filled Fe_A-e(↑) and Fe_B-a_1(↓) bands as well as a narrow V_B-t_{2g} band, corresponding to the V^{3+}/V^{2+} redox energy (the t_2^3 band) just overlapping the Fermi energy. Fast electron transfer ($\tau_h \approx 10^{-12}$ s) between the ions would define the significance of the formal valences shown as only an indication of the mean 3d-electron populations on the Fe_A, Fe_B and V_B atoms. Any overlap of the $V_B : t_2^3$ band with the Fermi energy should presumably increase with x since the $V_B : t_2^3$ bandwidth should increase with the density of V_B–V_B interactions; it should thus be a maximum for the end member $Fe[V_2]O_4$ where the valence distribution $Fe^{2+}[V_2^{3+}]O_4$ has been confirmed by Mössbauer measurements at 115 K, 137 K, and room-temperature[152]. This observation would seem to argue against a mixed valency for the ground state of vanadium.

The transport properties of the system $Fe[V_xFe_{2-x}]O_4$ exhibit a sharp cusp at $x \approx 1.89$[149]; and a large, positive Seebeck coefficient correlates well with a small hole concentration on the A sites in the domain $1.89 \lesssim x \leqslant 2.0$. The Mössbauer data also support the valency distribution

$$Fe^{2+}[Fe_{2-x}^{3+}V_x^{3+}]O_4 \quad \text{for} \quad 1.89 \lesssim x \leqslant 2.0 \tag{26}$$

If strong electron-phonon interactions tend to stabilize a mean Fe_B-atom valence of 2.5 + for $x < 1.89$, as is also indicated by the Mössbauer data ($\tau_h < 10^{-8}$ s) and Eq. (1), then the valence distribution would be

$$Fe_{1-0.5x}^{3+}Fe_{0.5x}^{2+}[Fe_{2-x}^{2.5+}V_x^{3+}]O_4 \quad \text{for} \quad 0 \leqslant x \lesssim 1.89 \tag{27}$$

and the rapid increase in the conductivity as (2 − x) increases to 1 would indicate that the conduction is carried primarily by a rapid charge transfer on the Fe_B atoms. A sharp drop in the Seebeck coefficient with increasing (1.89 − x), a change in sign occurring at $x \approx 1.4$, is not inconsistent with this model; but agreement would be improved by allowing for some oxidation and/or a fractional charge transfer to the vanadium atoms. For x = 1, a 10-percent V^{2+} concentration is consistent with the observed paramagnetic Curie constant $C_m = 8.60$ emu/mole[154].

At low temperatures ($T < T_c \lesssim 127$ K), a cooperative Jahn-Teller distortion to tetragonal symmetry, (c/a > 1) for $(2 - x) \leqslant 0.1$ and (c/a < 1) for $0.15 \leqslant (2 - x) \leqslant 0.5$, has been reported[149]. Both the Fe^{2+} and V^{3+} ions are Jahn-Teller ions in cubic crystalline fields, and a localized V^{3+}:$^3T_{1g}$ configuration capable of distorting to either sign, (c/a < 1) if spin-orbit coupling not quenched, can probably account for the data without requiring localized Fe^{2+} configurations[136].

Quadrivalent Solute. The quadrivalent ions Sn^{4+}, Ti^{4+} and Mo^{4+} each have octahedral-site preferences: Sn^{4+} is too large for an A site, Ti^{4+} and Mo^{4+} each have important d components to their covalent bonding with O^{2-} ions. Moreover, substitution of these ions into magnetite tends to stabilize charge transfer to the Fe_A sites, as does substitution of trivalent ions. In the cases of nonmagnetic Sn^{4+} and Ti^{4+} ions, it is possible to deduce the valence distribution from a measure of the low-temperature spontaneous magnetization and the assumption of Néel ferrimagnetism. In this manner, Basile et al.[155] have determined that y increases continuously with x, reaching a value y = 0.25 at x = 0.4, in the formulation

$$Fe^{3+}_{1-y}Fe^{2+}_{y}[Sn^{4+}_{x}Fe^{2+}_{1-y+x}Fe^{3+}_{1+y-2x}]O_4 \qquad (28)$$

The system has a solid-solution range $0 \leqslant x \leqslant 0.58$ at 750 °C[156–159]; the composition $SnFe_2O_4$ reported by Banerjee et al.[160] appears to be unstable at 750 °C. The complete range of *titanomagnetite* solid solutions, from magnetite (Fe_3O_4) to ulvöspinel (Fe_2TiO_4), is well known to mineralogists; and Banerjee et al.[161] have similarly deduced the valence distributions as

$$\begin{aligned}
&\text{(i)} \quad \text{for} \quad x \leqslant 0.2, && Fe^{3+}[Fe^{3+}_{1-2x}Fe^{2+}_{1+x}Ti^{4+}_{x}]O_4 \\
&\text{(ii)} \quad \text{for} \quad 0.2 \leqslant x \leqslant 0.8, && Fe^{3+}_{1.2-x}Fe^{2+}_{x-0.2}[Fe^{3+}_{0.8-x}Fe^{2+}_{1.2}Ti^{4+}_{x}]O_4 \qquad (29) \\
&\text{(iii)} \quad \text{for} \quad 0.8 \leqslant x \leqslant 1, && Fe^{3+}_{2-2x}Fe^{2+}_{2x-1}[Fe^{2+}_{2-x}Ti^{4+}_{x}]O_4
\end{aligned}$$

This assignment is intermediate between two previous assumptions[162, 163] for this system.

Comparison of the physical properties of these two systems begins with a consideration of the energies of the bottoms of the $Ti^{4+/3+}$: $3d^1$ and $Sn^{4+/3+}$: $5s^1$ bands relative to one another and to the $Fe^{3+/2+}$: $3d^6$ bands on A and B sites. A measure of the relative positions of the bottoms of the conduction bands in n-TiO_2 and n-SnO_2 can be obtained from their flat-band potentials in aqueous solution at the pH corresponding to their zero zeta potential. The $Sn^{4+/3+}$: $5s^1$ band edge lies at least 0.5 eV below the $Ti^{4+/3+}$: $3d^1$ band edge. Furthermore, the Sn^{4+} : 5s wavefunction has a greater radial extension than the Ti^{4+} : $3d^1$ wavefunctions. Therefore, we can unambiguously predict that mixing of charge-transfer excited states $Fe^{2+}_A + M^{4+} \rightarrow Fe^{3+}_A + M^{3+}$ will be significantly more important for M = Sn than for M = Ti. Moreover, although the valence states Fe^{2+} and Ti^{4+} are stabilized in ilmenite and the M^{4+}-ion bandwidth would be smaller in the spinels, nevertheless charge transfer from Fe^{2+} to Ti^{4+} ions in ilmenites only requires i.r. excitations, so the bottom of the $Sn^{4+/3+}$: $5s^1$ band in $Fe_{3-x}Sn_xO_4$ is energetically accessible for strong mixture of charge-transfer excited states.

In both $Fe_{3-x}Ti_xO_4$ and $Fe_{3-x}Sn_xO_4$, there is good evidence that fast electron transfer ($\tau_h < 10^{-8}$ s) breaks down in the compositional range $0.2 < x < 0.5$. In the system $Fe_{3-x}Ti_xO_4$, a dramatic increase with x in the coercivity of polycrystalline toroids occurs in the range $0.5 < x \leqslant 1$[161]; the most straightforward explanation is that it reflects a giant single-ion anisotropy associated with localized $^5T_{2g}$ configurations on the Fe^{2+}_B ions[40]. Since the concentration of Fe^{2+}_B ions remains constant in the range $0.2 \leqslant x \leqslant 0.8$, according to formulation (29), the hysteresis data strongly suggest an increase with (x − 0.2) in the concentration of localized $^5T_{2g}$ configurations at Fe_B^{2+} ions trapped near Ti^{4+} ions. Mössbauer data from 140 to 400 K for compositions $0.9 \leq x \leq 1$ confirm the presence of only localized configurations ($\tau_h > 10^{-8}$ s) at Fe_B^{2+} ions at these larger x[164]. In the system $Fe_{3-x}Sn_xO_4$, Mössbauer subspectra for both Fe^{2+} and Fe^{3+} ions on both A and B sites appear for x > 0.2, which provides direct evidence for a $\tau_h > 10^{-8}$ s in Sn^{4+}-rich regions.

The importance of charge-transfer excited states appears from an analysis of the magnetic data, including the paramagnetic susceptibility, on the basis of Néel's theory. Basile et al.[157] determined the evolution with x of the Fe_B–Fe_B and Fe_A–Fe_A interactions in the system $Fe_{3-x}Sn_xO_4$. They found that the strength of the ferromagnetic component of the Fe_B–Fe_B interactions increases systematically with the concentration of Fe_B^{2+} ions, and hence with the concentration of mobile $a_1(\downarrow)$ electrons for double exchange. In addition, they found an abrupt introduction of a ferromagnetic component to the Fe_A–Fe_A interactions with the appearance of Fe_A^{2+} ions. This unusual finding implies an important electron-transfer interaction across an Fe_A^{2+}–O–Sn–O–Fe_A^{2+} linkage. Indeed, Melzer et al.[165] have demonstrated an important $Fe^{2+} + Sn^{4+} \rightarrow Fe^{3+} + Sn^{3+}$ charge-transfer excited-state mixing into the Mössbauer spectrum of the ^{119}Sn signal in a double Mössbauer measurement with ^{57}Fe. The ^{119}Sn spectrum gave not only direct evidence of 3 d-electron density on the Sn, but also a six-line spectrum that changed sensitively with the concentration of Fe_A^{2+} ions.

There is disagreement concerning the valence state of molybdenum in the system $Fe_{3-x}Mo_xO_4$. In Fe_2MoO_4 some investigators[166–170] assign it the valence state Mo(IV); others[10, 171] the state Mo(III). This confusion arises because there is little, if any, splitting of the $Mo^{4+/3+}$: $4d^3$ and $Mo^{5+/4+}$: $4d^2$ bands in oxides occupied by Mo^{4+} ions sharing common octahedral-site edges. It is now well established that, in such a situation, the Mo-t_2 orbitals form either a narrow, itinerant-electron band or Mo–Mo cluster orbitals[4]. In the system $Fe_{3-x}Mo_xO_4$, the Mo-t_2 orbitals can be expected to form a narrow, partially filled band. The formal valence on the molybdenum depends upon the number of 4 d electrons per molybdenum in the band. If the Fermi energy E_F for the formal valence Mo^{4+}: $4t^2$ lies above the top of the $Fe^{3+/2+}$: $3d^6$ bands, the formal-valence assignment $Fe^{2+}[Fe^{2+}Mo^{4+}]O_4$ is appropriate for the end number, and the valence states in the system $Fe_{3-x}Mo_xO_4$ may include $Mo^{5+/4+}$ mixed valencies. Chemical support for such an assignment comes from the observation that $FeMoO_4$ has the formal valences Fe(II) and Mo(IV); but the perovskite Sr_2FeMoO_6 contains the formal valences Fe(III) and Mo(V)[172]. On the other hand, the spinel structure allows for Mo–Mo interactions across shared octahedral-site edges; these interactions lower the bonding states of the Mo-t_2 bands, but should leave the energy of the center of the Mo-t_2 bands above the top of the $Fe^{3+/2+}$: $3d^6$ bands. We may conclude that reduction of molybdenum all the way to Mo^{3+} is improbable, but placement of E_F for the formal valence Mo^{4+}: $4t^2$ relative to the top of the $Fe^{3+/2+}$: $3d^6$ bands remains ambiguous. If it lies below the top of the $Fe^{3+/2+}$: $3d^6$ bands, assignment of formal valences to the system $Fe_{3-x}Mo_xO_4$ becomes rather meaningless as rapid charge transfer will occur between all the cations.

Given this situation, it is particularly important to characterize the end member Fe_2MoO_4. It is a ferrimagnetic, inverse spinel $Fe[FeMo]O_4$ with a Curie temperature $T_c \simeq 348$ K and a small spontaneous ferrimagnetic moment at T = 0 K: $\mu_{mole} \simeq 0.18\ \mu_B$[166]. Were E_F above the top of the $Fe^{3+/2+}$: $3d^6$ bands on both A and B sites, the high conductivity and apparent lack of a spontaneous molybdenum moment could be attributed to a partially filled Mo-t_2 band corresponding to formal valence Mo^{4+}: $4t_2^2$. In this case, the small ferrimagnetic moment would represent different orbital-moment contributions from Fe_A^{2+} and Fe_B^{2+} ions plus a small molybdenum moment induced by the internal molecular fields produced by ordering of the Fe^{2+}-ion spins. Moreover, such an assignment should give rise to Mössbauer isomer shifts at the iron that correspond to Fe^{2+} as predicted from Eq. (1). In fact, a room-temperature Mössbauer shift of about 0.55 mm/s wrt elemental iron has been observed for both A and B sites, which can be distinguished by their different quadrupole splittings[10]. These results clearly point to an E_F that intersects the $Fe^{3+/2+}$: $3d^6$ bands on both A and B sites with rapid electron transfer between all the cations of the spinel. To talk of formal valence states under these conditions is probably misleading.

In the system $Mo_xFe_{3-x}O_4$, there is a regular variation with x, from Fe_3O_4 to $MoFe_2O_4$, in the lattice parameter (a_0), the magnetization (M_s), the Curie temperature (T_c) and the electrical conductivity (σ, ΔH_m); apparently substitution of Fe by Mo leads to a mixed iron valence on both sites for $0 < x < 1$. It is worth noting that ΔH_m varies smoothly with x from a $\Delta H_m < kT$ for $x = 0$ in the temperature interval $300 < T < 600$ K to 0.027 eV for $x = 1.0$, where the narrow minority-spin band is maximally perturbed and filled by the introduction of substitutional Mo[173].

Complete Substitution of Iron on B Sites

As previously stressed in paragraph 1.2.4., an important question is whether charge transfer on A sites proceeds via Fe_A–O–O–Fe_A or Fe_A–O–M_B–O–Fe_A transfer. Any Fe_A–Fe_A interaction is considered to be negligible for two reasons:

(i) the Fe_A–Fe_A distance is relatively large, ca. 3.5 Å as against ca. 3.0 Å for the Fe_B–Fe_B separation.

(ii) the transfer integral is nearly zero because the π-bonding e orbitals are directed along the $\langle 100 \rangle$ directions whereas the Fe_A–Fe_A axes are directed along $\langle 111 \rangle$ axes, which makes for negligible direct overlap of the e orbitals on neighboring Fe_A atoms.

In the previous section we had demonstrated that where the Fe_A^{2+} and acceptor M_B energies have overlapping energies, as in the case where $M_B = Sn_B$ or Mo_B, then an important charge transfer can occur via Fe_A–O–M_B–O–Fe_A interactions. We now ask whether significant Fe_A–O–O–Fe_A interactions can occur where the Fe_A–O–M_B–O–Fe_A interactions are weak. Two systems have recently been studied that relate to this question[174]:

$$Fe_{1-2x}^{2+}Fe_{2x}^{3+}\square_{x/4}[Cr_2^{3+}\square_{x/2}]O_{4+x} \quad \text{for} \quad 0 \leqslant x \leqslant 1/2$$

$$Fe_x^{2+}Fe_{1-x}^{3+}[Ni_{1-x}^{2+}Cr_{1+x}^{3+}]O_4 \quad \text{for} \quad 0 \leqslant x \leqslant 1$$

In these systems, the $Fe_A^{3+/2+}$: $3d^6$ energy lies within a large energy gap ($U_{eff} = U' + \Delta_c$) separating the $Ni^{2+/+}$: $3d^9$ and $Ni^{3+/2+}$: $3d^8$ energies, the $Cr^{3+/2+}$: $3d^4$ and $Cr^{4+/3+}$: $3d^3$ energies. In this situation, the Fe_A–O–M_B–O–Fe_A interaction is weak, and any significant charge transfer between A-site iron atoms would be via Fe_A–O–O–Fe_A pathways. At room temperature, the Mössbauer spectra clearly show distinguishable Fe_A^{2+} and Fe_A^{3+} signals, which is indicative of slow ($\tau_h > 10^{-8}$ s) electron transfer[85]. The origin of the mixed-valence signal recorded at higher (T > 300 °C) temperatures is not established; there is no need to associate it with Fe_A–O–O–Fe_A interactions.

This finding is not surprising; it has been anticipated in the small magnitudes of the Fe_A–O–O–Fe_A superexchange interactions. It follows from the following two considerations:

(i) the O–O interactions are relatively weak
(ii) the AO_4 tetrahedra have no common ion, so even in concentrated systems polaron formation is not inhibited by polaron-polaron interactions.

2.4.4 Substitution on Both A and B Sites

In many spinel systems, substitution for iron may result in a distribution of the solute on both A and B sites that varies strongly with temperature. In this case it is possible to vary the physical properties by quenching from different annealing temperatures. In each of the examples chosen to illustrate this effect, there is a point of additional interest to be introduced.

(a) *$Mg_xFe_{3-x}O_4$*. Magnesium ferrite has the cation distribution[175–177]

$$Fe^{3+}_{1-\alpha}Mg^{2+}_{\alpha}[Mg^{2+}_{1-\alpha}Fe^{3+}_{1-\alpha}]O_4 \tag{30}$$

with

$$\alpha(1 + \alpha) = (1 - \alpha)^2 \exp(-W/kT)$$

A $W/k \simeq 1345$ K was obtained by Tellier[175], who used a slow cooling rate (10 °C/hr) to approach equilibrium. Since $MgFe_2O_4$ is a Néel ferrimagnet, the saturation magnetization at T = 0 K and the Curie temperature vary sensitively with α.

In such a compound, a mixed $Fe^{3+/2+}$ valence is only introduced through impurities and/or oxygen vacancies. Upon reduction, the mobile electrons become trapped at oxygen vacancies; the activation energies for conduction are 0.48 eV in the ferrimagnetic domain and 0.57 eV in the paramagnetic domain[178].

In the system $Mg_xFe_{3-x}O_4$, on the other hand, an $x \neq 1$ introduces a mixed $Fe^{3+/2+}$ valence; and Dormann et al.[179] have explored the evolution of the Mössbauer spectra from 4.2 K to 300 K over the compositional range $0 \leq x \leq 0.4$. Three temperature ranges were distinguished: $T < T_p$, where T_p is the transition temperature to a less symmetric phase, $T_p < T < T_e$, and $T_e < T$, where T_e is a temperature above which the mean isomer shift for the Fe_B ions correspond to mean charges diverging from 2.5 +. The temperature T_p decreases with increasing x from $T_V \simeq 120$ K at x = 0 to about 20 K at x = 0.4. The temperature T_e increases with x from T_V and becomes less distinct as it approaches 300 K near x = 0.4. In the interval $T_p < T < T_e$, the Fe^{3+}_A and Fe^{3+}_B ions give an unresolved sextet whose intensity increases with x, and the spectrum corresponds to Mg-rich domains with distinguishable Fe^{3+}_B and Fe^{2+}_B ions together with Mg-poor domains with a mean valence $Fe^{2.5+}_B$. These data support the idea that below 300 K strong electron-phonon coupling stabilizes the mixed-valence ratio $Fe^{3+}_B/Fe^{2+}_B \simeq 1$ where it can and small polarons where it cannot; it also supports a model of electron hopping everywhere between iron atoms at temperatures above 300 K even though a $\Delta H_m < kT$ may make the conductivity nearly temperature-independent.

(b) *$Co_xFe_{3-x}O_4$*. The system $Co_xFe_{3-x}O_4$, $0 \leqslant x \leqslant 1$, contains Co^{2+} ions[180] on both octahedral and tetrahedral sites[181, 182]. A Co_A^{2+} ion has the localized configuration $^4A_{2g}$, which is nondegenerate with a quenched orbital angular momentum; a Co_B^{2+} ion has a $^4T_{1g}$ configuration that imparts, below a Néel-ferrimagnetic Curie temperature, a giant single-ion anisotropy and magnetostriction. Therefore the Co^{2+}-ion distribution influences sensitively not only the spontaneous ferrimagnetic magnetization and Curie temperature, but also the magnetic anisotropy and magnetostriction[181–183].

In this system, $CoFe_2O_4$ is not the end member; it is possible to go all the way to Co_3O_4 because of the possibility of introducing holes into the $Co_B^{3+/2+}$: $3d^7$ band. Jonker[184] has provided an elegant demonstration of how the Seebeck coefficient α changes sign from a high negative value to a high positive value as x passes through $x = 1$, see Fig. 11. As x passes through $x = 1$, the Fermi energy E_F drops out of the bottom of the $Fe_B^{3+/2+}$: $3d^6$ band through an energy gap E_g into the $Co_B^{3+/2+}$: $3d^7$ band. The maximum magnitudes of the positive and negative Seebeck coefficients increase and the compositional separation between them decreases as E_g increases in accordance with Eq. (15).

Room-temperature Mössbauer data[182, 185, 186] for the system $Co_xFe_{3-x}O_4$ indicate statistical distributions of Co^{2+} ions within A and B sites and a fast electron transfer ($\tau_h < 10^{-8}$ s) on the iron for all $x < 1.0$.

(c) *$Mn_xFe_{3-x}O_4$*. The system $Mn_xFe_{3-x}O_4$ has the atomic distribution

$$Mn_{1-\varepsilon}Fe_\varepsilon[Mn_{x+\varepsilon-1}Fe_{3-\varepsilon-x}]O_4 \tag{31}$$

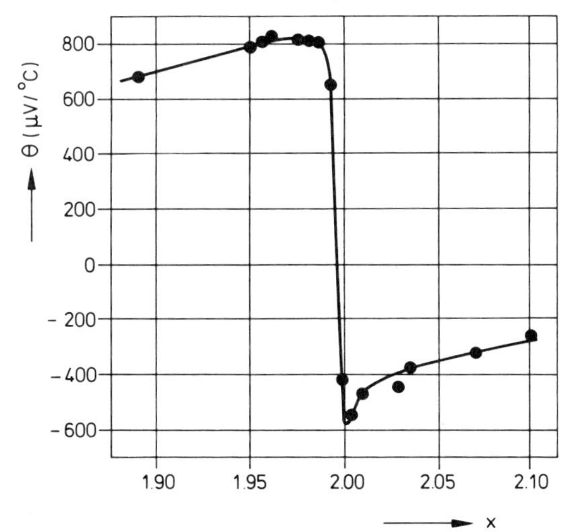

Fig. 11. Thermoelectric power versus composition for $Fe_{3-x}Co_xO_4$, after[184]

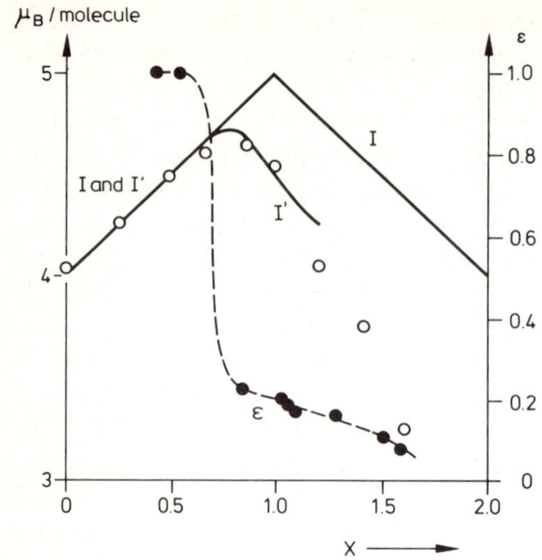

Fig. 12. Comparison of Néel theory (*curve I*) with (*open circles*) measured Bohr magneton number $n_B = \bar{\mu}_{mole}/\mu_B$, extrapolated to T = 0 K, for $Mn_xFe_{3-x}O_4$, and (*black circles*) atomic-distribution parameter ε in Eq. (31), after[191)]

where the parameter ε varies with x as shown in Fig. 12[187)]. The spontaneous magnetization, in μ_B per molecule, for Néel ferrimagnetism (curve I) is compared with experiment (open circles[188)] in this same figure). In this case a deviation from the Néel model of curve I, which is based on the assumption that

$$Mn^{3+} + Fe^{2+} \rightarrow Mn^{2+} + Fe^{3+} \tag{32}$$

on both the A and B sites, is seen to occur for x larger than a critical value beyond which ε decreases sharply. Simsa[189)] interpreted this deviation to be due to the coexistence of Mn^{3+}_B and Fe^{2+}_B ions; but he accepted that reaction (32) holds for the A sites. Lotgering[190)], on the other hand, assumed reaction (32) must hold for both A and B sites and therefore that a small deviation from collinear B-site spins is associated with Mn^{2+}_B ions having six nearest-neighbor Mn^{2+}_A ions. The Lotgering model conforms to the observation that substituting Mn for iron in a variety of oxides quenches the dielectric losses associated with small Fe^{2+}/Fe^{3+} ratios in slightly reduced Fe(III) oxides, see the section on $Li_{0.5}Fe_{2.5}O_4$. With Mössbauer measurements on paramagnetic $Mn_{1-\varepsilon}Fe_{\varepsilon}$-$[Fe_{0.9-\varepsilon}Mn_{0.8+\varepsilon}Ti_{0.3}]O_4$, $Zn^{2+}[Fe_{0.2}Mn_{0.8}Ga_{0.8}Ti_{0.2}]O_4$, $Zn^{2+}[Fe^{2+}_{0.4}Ga^{3+}_{1.2}Ti^{4+}_{0.4}]O_4$, $Zn^{2+}[Fe_{1.5}Mn_{0.5}]O_4$ and $Zn^{2+}[Fe^{3+}_{1.5}Fe^{2+}_{0.25}Ti^{4+}_{0.25}]O_4$, Lotgering and Van Diepen[191)] were able to demonstrate convincingly that reaction (32) holds for the B sites as well as the A sites.

As x increases through x = 1, the Fermi energy drops from the bottom of the $Fe^{3+/2+}$: $3d^6$ bands to the top of the $Mn^{3+/2+}$: $3d^5$ band, and the activation energy of the conductivity increases abruptly from 0.05 eV for small-polaron Fe^{2+}_B ions to 0.3 eV for small-polaron Mn^{3+} holes[191)]. The Mn^{3+}_B ion is a strong Jahn-Teller ion (configuration ^5E_g), so the holes in the $Mn^{3+/2+}$: $3d^5$ band form more stable small polarons.

(d) *$Cu_xFe_{3-x}O_4$*. The system $Cu_xFe_{3-x}O_4$ has been investigated primarily in the range $0.5 < x < 1.0$, which can be expected to contain the mixed valences $Cu^{2+/+}$ because of the reaction

$$Fe^{2+} + Cu^{2+} \rightarrow Fe^{3+} + Cu^{+} \tag{33}$$

Whereas the Cu^{2+} ion does not compete strongly with Fe^{3+} ions for the A sites, the Cu^{+} ion does. Moreover, the Cu^{+} ion can be fairly mobile in close-packed anion arrays, and we will suggest that accommodation of Cu^{+} in interstitial positions may be a source of some confusion about valence states. Therefore, let us consider first what evidence there is for the possibility of interstitial ions associated with the spinel structure.

The spinel structure is incompatible with a significant population of interstitial ions. Placement of a cation in the empty 16c octahedral sites of the cubic-spinel space group Fd3m is inhibited by the strong electrostatic forces arising from the two A-site cations with which a 16c site shares common faces. However, it is now known that lithium can be inserted either chemically or electrochemically at room temperature into mangano- and ferrospinels without perturbing the $[B_2]X_4$ framework[192)]. The lithium enter the interstitial 16c sites as Li^{+} ions and donate an electron to the host matrix, reducing Fe^{3+} to Fe^{2+} or Mn^{3+} to Mn^{2+} or Mn^{4+} to Mn^{3+}, for example. The electrostatic force between the Li^{+} ions and the host A-site cation appears to displace the A-site cation to positions 16c, so the product has cations only on octahedral sites of a close-packed oxide-ion array; but a peculiar cation ordering persists: the octahedral 16d sites of the cubic-spinel space group Fd3m remain occupied as in the original spinel phase, and the Li^{+} and originally A-site cations are randomly distributed with any vacancies on the 16c sites. We may expect Cu^{+} ions to be similar to Li^{+} ions in their ability to occupy the interstitial 16 sites.

With this background, it is instructive to consider the end member $CuFe_2O_4$, which exhibits a cooperative Jahn-Teller distortion to tetragonal ($c/a \simeq 1.05$) symmetry at room temperature if prepared as stoichiometric $Fe^{3+}[Fe^{3+}Cu^{2+}]O_4$. The low-temperature distortion depends on a Cu_B^{2+} concentration exceeding a critical value. If the compound is heated in air well above the tetragonal-cubic transition temperature $T_t = 370\,°C$ and quenched, the cubic phase becomes stabilized to lowest temperatures. Since a Jahn-Teller phase transition is diffusionless, this observation indicates that the concentration of Cu_B^{2+} ions has been reduced by heating in air, and the accepted interpretation is that Cu^{2+} ions have been transferred to A sites in a manner analogous to what happens to Mg^{2+} ions in $MgFe_2O_4$. However, this interpretation appears to be too naive.

A Cu_B^{2+} ion has a localized-electron configuration ^2E_g, which makes it a strong Jahn-Teller ion[193)]. However, the Cu_A^{2+} ion has the localized-electron configuration $^2T_{2g}$, which makes it also a Jahn-Teller ion; and even in dilute concentrations it is capable of a cooperative distortion to tetragonal ($c/a > 1$) symmetry below a collinear-spin magnetic-ordering temperature T_c[40, 194)], as has been demonstrated experimentally in the system $Cu_{1-t}^{2+}Fe_t^{3+}[Cr_{2-t}^{3+}Cu_t^{2+}]O_4$[195)]. Since $CuFe_2O_4$ is a Néel ferrimagnet, we must conclude that either the Cu_A^{2+} ions in quenched $CuFe_2O_4$ do not carry a localized-electron configuration, which seems unlikely, or the system has been reduced at high temperatures, which is more probable. Given reaction (33), reduction should create Cu^{+} ions, and these could occupy interstitial sites as well as A sites:

$$(1 + \alpha)Fe^{3+}[Fe^{3+}Cu^{2+}]O_4 \rightarrow \{Cu_{4\alpha}^{+}\}_{16c}Fe_{1-5\alpha}^{3+}Cu_{4\alpha}^{+}\square_{\alpha}[Fe_{1+7\alpha}^{3+}Cu_{1-7\alpha}^{2+}]O_4 + 1/2\,\alpha O_2 \qquad (34)$$

where it is assumed that charge neutrality and electrostatic forces would stabilize about an A-site vacancy a large Cu^{+}-ion cluster, $\square Cu_8^{+}$, in which half the Cu^{+} ions of the cluster occupy the four 16c octahedral sites neighboring the A-site vacancy, and half occupy the A sites that share octahedral-site faces with the Cu^{+}-occupied 16c sites. Such a cluster would be stable. Such clusters should be detectable with neutron-scattering experiments.

The other end member $Cu_{0.5}Fe_{2.5}O_4$ is also a Néel ferrimagnet with a Curie temperature T_c and saturation magnetization μ_{mole} that varies with the thermal history of the samples, and a number of studies have been made of the cation and valency distribution in the compound[196–211]. A polycrystalline sample quenched from 1150 °C has been characterized by several techniques[208]. This sample had a cubic room-temperature lattice parameter $a_0 = 8.424 \pm 0.002$ Å, an oxygen parameter $u = 0.385$, a ferrimagnetic Curie temperature $T_c = 348 \pm 2$°C, and a saturation magnetization per molecule $\mu_{mole} = 5.06\ \mu_B$. According to the most extensive and recent investigation of Hannoyer[212], the valence and atomic distribution would be:

$$Cu^{+}_{0.1}Cu^{2+}_{0.2}Fe^{3+}_{0.7}[Cu^{+}_{0.1}Cu^{2+}_{0.1}Fe^{2+}_{0.3}Fe^{3+}_{1.5}]O_4$$

which is consistent with all the above given data and is also supported by the values of X absorption shifts, $X(K_{\beta 5})$ emission and XPS experiments. In this last technique care has been taken to check the ratio Cu/Fe at the surface in order to avoid any change due to a possible segregation such as the one observed recently by Brabers[213]. However, this formulation implies that the $Cu^{2+/+}$: $3d^{10}$ bands on Cu_A and Cu_B atoms are both broad and overlap the $Fe_B^{3+/2+}$: $3d^6$ band, which seems improbable. As the model is based on the constraint that all cations occupy spinel positions, it would seem physically more reasonable to relax the constraint. Therefore we suspect a chemical inhomogeneity may be present with Fe^{2+} ions trapped in cation-rich domains

$$\{Cu^{+}_{4\alpha}\}_{16c}Cu^{+}_{4\alpha}\square_{\alpha}Fe^{3+}_{1-5\alpha}[Fe^{3+}_{1+7\alpha}Fe^{2+}_{1-7\alpha}]O_4$$

that are segregated from cation-deficient regions

$$Fe^{3+}[Fe^{3+}_{1.5}Cu^{2+}_{0.25}\square_{0.25}]O_4$$

2.4.5 *Substitution of Two Different Cations*

Two different cations may be substituted into Fe_3O_4 for a variety of purposes. The experiments of Lotgering and Van Diepen[214], for example, provided an interesting example of how judicious substitutions could be used to establish reaction (32) for B-site ions, and Mn substitutions into $LiFe_5O_8$ to suppress dielectric loss illustrates the use of a second-cation substitution to optimize a material for a specific application. In this final section on spinels, we choose to discuss the solid solution between $Zn[Fe_2]O_4$ and $Ge[Fe_2]O_4$, which would appear to offer the opportunity to study the evolution of the character of the B-site charge carriers with Fe_B^{2+}-ion concentration. Ideally, the system would be $Zn^{2+}_{1-x}Ge^{4+}_{x}[Fe^{3+}_{2-2x}Fe^{2+}_{2+2x}]O_4$.

The cubic ($a_0 = 8.441$ Å) end member $Zn[Fe_2]O_4$ was seen to be n-type with mobile electrons moving at the bottom of a narrow $Fe^{3+/2+}$: $3d^6$ band. Transport properties intermediate between those expected for small polarons or for itinerant electrons suggest a diffusional motion, but with a motional enthalpy $\Delta H_m < kT$. In the cubic ($a_0 = 8.411$ Å) end member $Ge[Fe_2]O_4$, the transport and optical properties of high-purity, p-type single crystals – grown by chemical-vapor ($TeCl_4$) deposition – indicate that the mobile holes are thermally excited from acceptor sites into a narrow, filled $Fe^{3+/2+}$: $3d^6$

valence band of itinerant-electron states[215]. The acceptor ionization energy is 0.39 eV; it is probably associated with cation vacancies. Although a temperature-independent ($200 < T < 320$ K) Hall mobility of 2 cm^2/Vs is within the itinerant-electron range, see Eq. (9), it is necessary to assume an important correlation splitting U′ of the $Fe^{2+/+}$: $3d^7$ and $Fe^{3+/2+}$: $3d^6$ bands. Also, the compound is paramagnetic in the temperature range of measurement, and the theory of the Hall effect uses a spin-dependent transfer energy. Therefore some doubt is cast on the quantitative significance of the Hall mobility. Nevertheless, we accept it as given, but we may assume that the sharp optical absorption edge at an $E_g \simeq 2.3$ eV gives the separation of the narrow Fe^{2+}: $3d^6$ valence band from the 4s band. Correlation splitting should make the gap between the $Fe^{2+/+}$: $3d^7$ band and the $Fe^{3+/2+}$: $3d^6$ band larger than 2.3 eV. Mössbauer, magnetic and optical data give a trigonal component to the crystal-field splitting of about 1000 cm^{-1}[215–217]. An overlap of the e and a_1 bands of t_2 parentage would give holes at the top of the $Fe^{3+/2+}$: $3d^6$ band access to all the orbitals of t_2 parentage whereas electrons at the bottom of the $Fe^{3+/2+}$: $3d^6$ band would tend to be restricted to states of a_1 parentage near the center of the Brillouin zone.

Unfortunately it appears difficult to prepare the ideal mixed system. Miyahara and Sai[218] observed large saturation magnetizations and high ferrimagnetic Curie temperatures indicative of the presence of Fe_A–O–Fe_B interactions. They used X-ray and magnetization data to estimate that 30 percent of the A-sites are Fe^{3+} in $Zn_{0.5}Ge_{0.5}Fe_2O_4$. A study by Robbins[219] has also indicated that Fe^{3+} ions will tend to displace Ge^{4+} ions on A-sites.

Subsequently Grandjean and Gérard[220] investigated compositions with $x = 0.25$, 0.5, and 0.75 prepared in evacuated silica tubes at 950 °C. Within experimental error, they could not identify in their Mössbauer spectra (from 77 to 623 K) any A-site iron in their preparations. They also found Curie temperatures that rose to a maximum $T_c \simeq 250$ K for $x = 0.5$, as might be expected for a ferromagnetic Fe_B–Fe_B double-exchange interaction due to mobile electrons in a narrow band of itinerant-electron states. They also reported a spectral broadening at higher temperatures that they attributed to excitation into an unspecified conduction band.

More recently, reinvestigation[221] of the $Zn_{1-x}Ge_xFe_2O_4$ system has led to the conclusion, based upon neutron-diffraction and magnetic measurements, that in $Zn_{0.5}Ge_{0.5}Fe_2O_4$ around 0.25 Fe^{3+} is located on A-sites, hence in rough agreement with Miyahara and Sai[218]. Moreover, it has been shown that the properties are not very sensitive to the thermal history of the samples. In view of the sensitivity to counter cations of the energy Δ splitting $Fe_A^{3+/2+}$ and $Fe_B^{3+/2+}$ redox energies, it would appear that the spectral broadening reported by Grandjean and Gérard[220] might well represent thermal excitation of the mobile electrons to Fe_A^{3+} ions having an isomer shift overlapping that of the Fe_B^{3+} ions.

Support for this tentative interpretation comes from a study by Tricker et al.[222] on the system $Zn^{2+}[Zn^{2+}_{(1-x)/2}Sn^{4+}_{(1+x)/2}Fe^{3+}_{1-x}Fe^{2+}_x]$ in which they also find evidence from Mössbauer data for thermal excitation of the mobile ferrous electron to an unspecified conduction band. In this case the conduction band is probably associated with the Sn_B^{4+} ions, which would have a 5s band edge close to the $Fe_B^{3+/2+}$: $3d^6$ redox energy.

3 Hexagonal Ferrites

3.1 Introduction

Since the discovery in 1951 that magnetoplumbite, $BaFe_{12}O_{19}$ is a hard, semiconducting ferrigmagnet of considerable technical interest, several other iron oxides with closely related hexagonal structures have been synthesized and studied. Of particular interest for this review are those containing mixed $Fe^{3+/2+}$ valency.

The magnetoplumbite structure, designated M, contains 10 close-packed layers on traversing the unit cell along the c-axis; four oxygen layers alternate with an ordered BaO_3 layer. The iron are distributed in oxygen octahedral, tetrahedral, and trigonal-bipyramidal sites in such a way as to give alternating spinel (S) and hexagonal (R) blocks along the c-axis as shown in Fig. 13. The BaO_3 layers contain the trigonal-bipyramidal iron; so the stacking sequence RSR*S*, where the asterisk designates inversion of the block, gives five distinguishable iron sites.

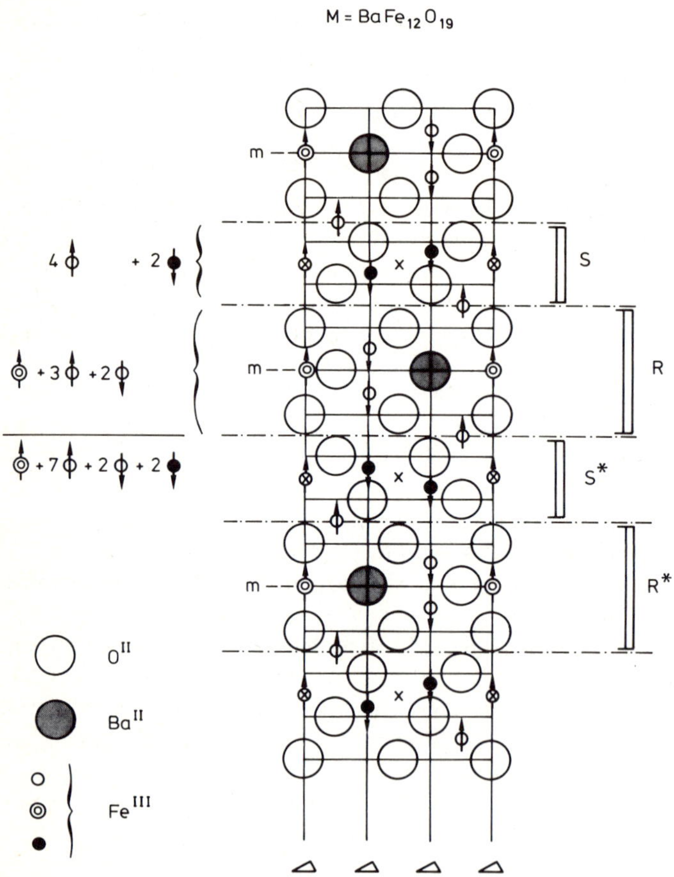

Fig. 13. Projection of the magnetoplumbite structure M showing spinel (S) and R blocks, the iron-spin orientations ($T < T_c$), and vertical lines of threefold symmetry parallel to c-axis, after[149)]

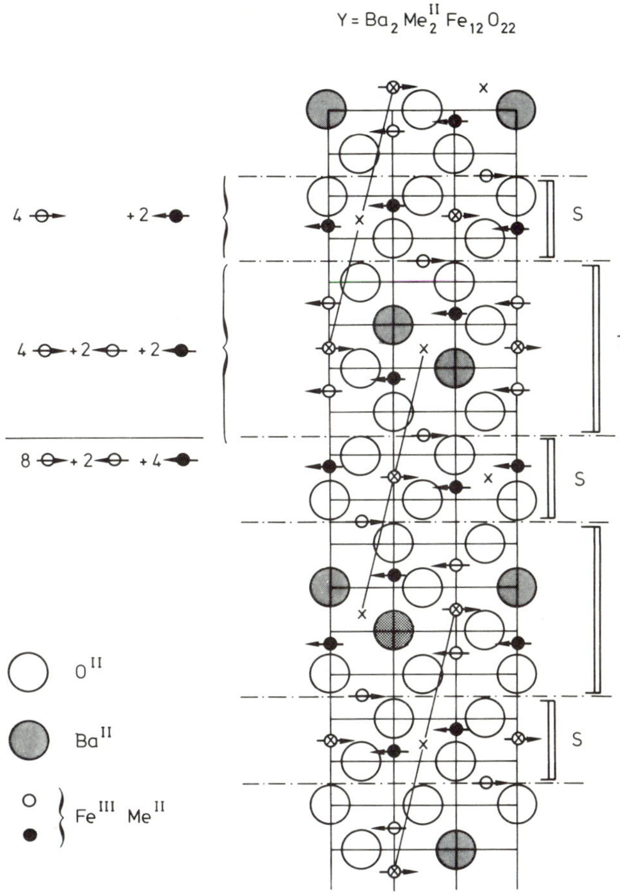

Fig. 14. Projection of the Y structure, as in Fig. 13, showing the T blocks, after[149)]

The Y structure of $Ba_2Fe_{14}O_{22}$ contains a double BO_3 layer, which forms a T block, interleaved with the same S blocks in the stacking sequence $(TS)_3$, see Fig. 14. The T block contains only octahedral and tetrahedral iron.

The family of structurally related hexagonal ferrites contain intergrowths of these elemental building blocks: S, R and T.

Another hexagonal ferrite of interest is $KFe_{11}O_{17}$, which has a structure similar to that of magnetoplumbite except for the $FeBaO_3$ layers, which are replaced by KO layers, see Fig. 15.

3.2 The W Structure

The ideal W structure of $BaFe_{18}O_{27}$ and $SrFe_{18}O_{27}$ contains S blocks twice as thick between each R block to give the stacking sequence $RS_2R^*S_2^*$. There are therefore seven non-equivalent iron sites, and Mössbauer spectra at 5 and 78 K were able to resolve all

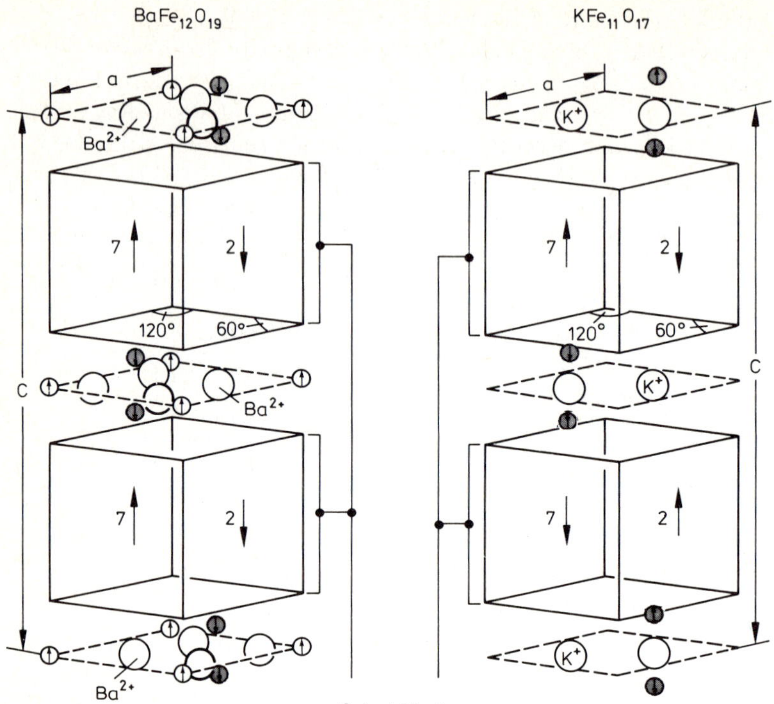

Fig. 15. Schematic comparison of unit cells of $BaFe_{12}O_{19}$ and $KFe_{11}O_{17}$, after[149)]

seven[223)]. The two Fe^{2+} ions per formula unit (four per unit cell) were shown to occupy two thirds of the octahedral sites 6 g, which are in the spinel slabs midway between the R blocks. This location is dictated by the requirements of local charge neutrality since the R block of the magnetoplumbite is $(BaFe_6O_{11})^{2-}$ and the spinel block is $(Fe_6O_8)^{2+}$. Moreover, the 6 g sites are ferromagnetically aligned, so any double-exchange coupling by the mobile electrons on the 6 g sites is cooperative with the overall superexchange coupling. At 78 K the hopping time is fast ($\tau_h < 10^{-8}$ s), but at 5 K an identifyable Fe^{2+} signal shows that $\nu_h < 10^8$ Hz is slow at lowest temperatures. These data are compatible with a diffusional motion for the mobile electrons, but with a small ΔH_m. The low Fe^{2+}/Fe^{3+} ratio stabilizes simple small polarons that may be ordered by electrostatic forces between them.

3.3 The M Structure Containing Fe(II) Ions

Substitution of La^{3+} for Ba^{2+} in magnetoplumbite introduces Fe^{2+} ions into the M structure. The end member has the approximate composition $LaFe^{2+}Fe^{3+}_{11}O_{19}$, and the mobile electrons are again forced by the electrostatics of the structure to occupy the octahedral sites at the center of the spinel slabs. These are the 2 a positions, which are completely filled at low temperatures with two Fe^{2+} ions per unit cell. At higher temperatures, thermal excitation to neighboring octahedral sites of the spinel block occurs[224–228].

Whereas sodium β-alumina tends to incorporate excess soda into its NaO layers to stabilize $Na_{1+2x}Al_{11}O_{17+x}$, excess potassium may be intercalated into the loosely packed layers with charge compensation occuring via iron reduction to form $K_{1+x}Fe_{11}O_{17}$[229]. Attempts to develop a reversible K^+-ion electrode with this material[230–233] have established that

(i) some interstitial O^{2-} ions are also present, as in the β-aluminas, to give the formula $K_{1.09+x}Fe^{3+}_{11-x}Fe^{2+}_{x}O_{17.05}$ for $x < 0.86$
(ii) the electronic conductivity, which increases with x, is diffusional and only weakly coupled to the ion flux, and
(iii) an ordering of K^+ ions appears to take place at the approximate composition $K_{1.7}Fe_{11}O_{17}$.

4 Wüstite and Substituted Wüstite

4.1 *Wüstite: $Fe_{1-x}O$*

Wüstite has great technical importance as well as academic interest. For example, it is the dominant phase in the blast furnace because it is reduced with more difficulty than are the starting ores, and the complex scales produced in hot corrosion have a large component of $Fe_{1-x}O$ in contact with the metal. Academically, the instability of stoichiometric FeO, except perhaps under high pressure[234], presents an intriguing theoretical problem. Moreover, the large solid-solution range of $Fe_{1-x}O$ above 570 °C and its disproportionation into elemental iron and magnetite in the interval $200 < T < 570$ °C are unusual properties for a monoxide with the rocksalt structure. Quenching from above 570 °C to room temperature allows study of $Fe_{1-x}O$ at low temperatures.

Density measurements[235] clearly establish that the system contains a close-packed anion array with cation vacancies. Moreover, neutron-diffraction data on samples quenched from high temperature have identified the presence of tetrahedral-site iron interstitials surrounded by four octahedral-site vacancies[236], but the nature of the elementary cluster associated with such an interstitial has been more difficult to establish[237–240]. Some of the clusters considered are shown in Fig. 16. From powder neutron diffraction at 985 and 1075 °C, Gavarri et al.[239] have found a ratio of octahedral-site vacancies to interstitial iron to be

$$\varrho = (x + i)/i = 2.4 \pm 0.4 \tag{36}$$

for all values of x, which suggests a single structural principle. This result does not establish the structural building block uniquely, but it does seem to eliminate several alternatives such as the Koch-Cohen cluster of Fig. 16g and the isolated 4:1 unit of Fig. 16a. The 8:3 cluster of Fig. 16c would satisfy Eq. (36) and carry a net negative charge, thereby satisfying the requirement of local charge neutrality.

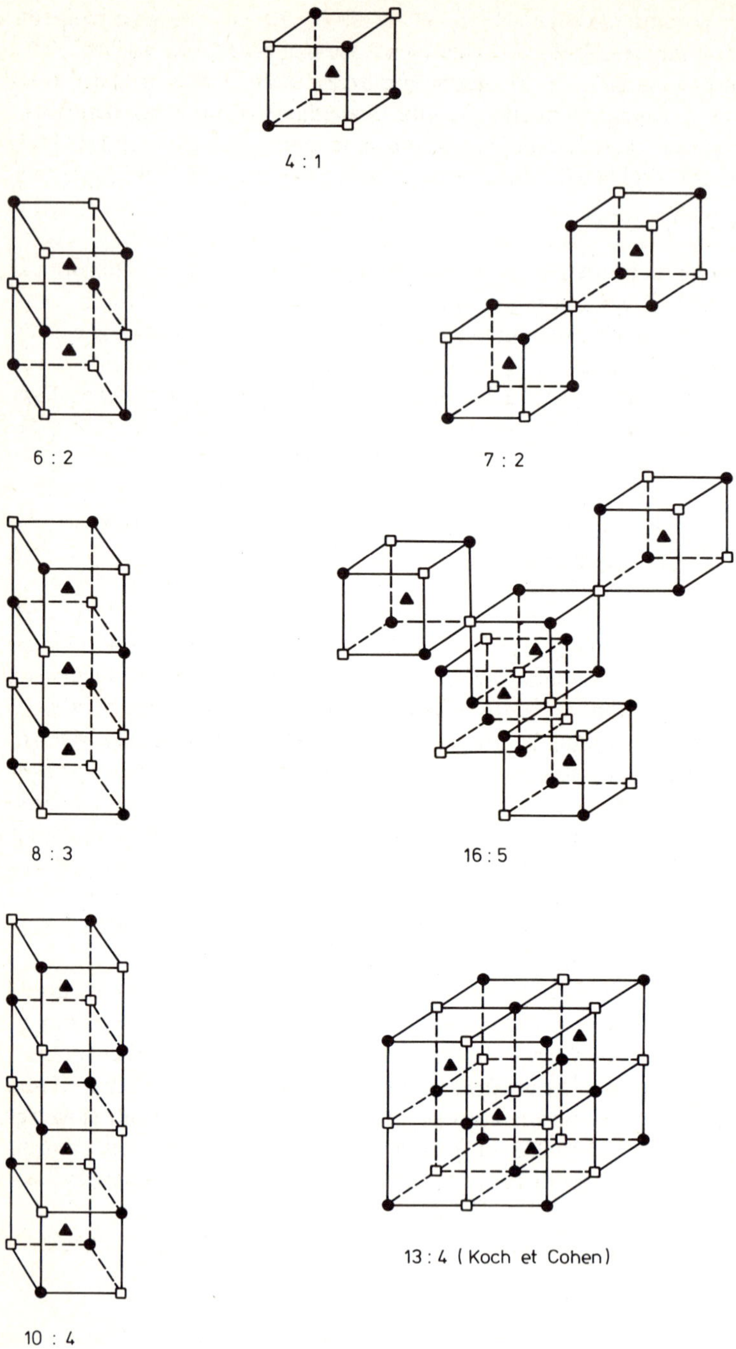

Fig. 16. Different clusters of Fe^{3+} interstitials and nearest-neighbor Fe vacancies considered for $Fe_{1-x}O$, after[239)]

Wüstite, $Fe_{1-x}O$, is antiferromagnetic below ca. 195 K, the exact Néel temperature depending upon x[241]. Neutron-diffraction measurements obtained below T_N[236] have confirmed a distortion from cubic to rhombohedral ($\alpha < 60°$) symmetry in the antiferromagnetic phase; they have also shown that the antiferromagnetic order consists of ferromagnetic (111) planes coupled antiparallel on progressing along the unique [111] axis of the rhombohedral cell, the iron moments being directed parallel to the [111] axis. Since dipole-dipole forces would orient the iron moments within the (111) planes, the structure and magnetic order clearly demonstrates that a cooperative Jahn-Teller distortion below T_N preserves the orbital angular momentum of a $^5T_{2g}$ localized configuration on Fe^{2+} ions[40]; the Jahn-Teller distortion and giant magnetic anisotropy is a sure diagnostic that the minority-spin electrons are localized.

On the other hand, the neutron data also show that the magnitude of the iron moments below T_N is considerably reduced from its theoretical value. Battle and Cheetham[242] recognized that the reduction in moment must be associated with the clusters, which attract the Fe^{3+} ions in order to preserve local charge neutrality. They proposed originally that ferrimagnetic clusters were decoupled from the matrix. However, on the basis of Mössbauer evidence for no magnetic coupling of the tetrahedral-site iron moments below T_N and of polarized-neutron data, Wilkinson et al.[243] have prepared a more satisfactory model in which the octahedral-site spins are all strongly coupled to one another, but dipole-dipole forces acting to orient the atomic moment in the (111) planes cause the spins to tilt away from the [111] axis. Only the component of the moment parallel to the [111] axis has long-range order. In the vicinity of the clusters, where the iron atoms are primarily Fe^{3+}, the moment is oriented about 35° from the (111) plane; between the defects it is about 30° from [111]. We may suspect some decoupling of S and L at the Fe^{2+} ions; the Russell-Saunders coupling would be weakened by superexchange interactions between electrons carrying an orbital angular momentum. The orbital contribution to the moment would be directed along the [111] axis as dictated by the Jahn-Teller distortion. The interstitial iron, in this model, are "frustrated" by the antiferromagnetic ordering of nearest-neighbor iron; their internal fields tend to cancel one another at the tetrahedral position, so the tetrahedral-site iron remains paramagnetic.

The room-temperature Mössbauer data show two broad peaks, and at least seven different interpretations have been suggested[244–248]. At 800 °C, a single broad peak indicates fast charge transfer between all the iron, including the interstitial Fe^{3+} ions[247].

The high-temperature transport data has been well rationalized[51] on the basis of a diffusional model with $\Delta H_m = 0.16$ eV for all x, which is an agreement with localized Fe^{2+}-ion configurations responsible for a cooperative Jahn-Teller distortion below T_N. The ratio $(1-c)/c$ obtained from Eq. (15) for the Seebeck coefficient, with a spin-degeneracy factor $\beta = 2$, gives the factor $c(1-c)$ entering the conductivity expression

$$\sigma = \gamma(Ne^2/kT)c(1-c)a_0^2\nu_0\exp(-\Delta H_m/kT) \tag{37}$$

derived from Eq. (4), (5), and (10). The conductivity expression varies with the partial pressure of oxygen as the $c(1-c)$ so obtained. Moreover, the factor $\beta = 2$ makes the Seebeck coefficient α change sign at $(1-c) = 1/3$ rather than 1/2. Actually α changes sign at $(1-c) = 1/4$ if the total Fe^{3+} and Fe^{2+} populations are included. Removal of the interstitial Fe^{3+} ions would only increase the discrepancy. However, confinement of the mobile Fe^{3+} ions to the neighborhood of the clusters can account qualitatively for a shift in the $\alpha = 0$ concentration to lower values of $(1 - c)$. Although the percolation limit must be exceeded to give good electronic transport, pockets of pure FeO matrix would exist in the space interstitial to the regions influenced by the clusters. The sites in these pockets do not belong to the total population of N sites available to the mobile holes in $(1 - c) = [Fe^{3+}]\,N$. Numerous earlier attempts to account for σ and α separately at constant x were less satisfactory. It is essential to introduce the factor β and to analyze σ at constant α to obtain a satisfactory evaluation of the effective factor $c(1 - c)$.

4.2 *Wüstite Solid Solutions*

The solid solutions of wüstite have been extensively studied and are reviewed elsewhere[249, 250]. The following facts are worth repeating here:

(i) Weak substitutions of Ti for Fe decrease the conductivity of $Fe_{1-x}O$ twice as efficiently as Cr substitutions. Since Cr^{3+} ions remove one hole from the $Fe^{3+/2+}$: $3d^6$ band and Ti^{4+} ions remove two, the concentration of mobile holes must be decreased twice as much with Ti substitutions as with a comparable Cr substitution. From the valence distribution $Fe^{2+}Ti^{4+}O_3$, we expect the $Ti^{4+/3+}$: $3d^1$ band to lie above the $Fe^{3+/2+}$: $3d^6$ band.

(ii) The Mg^{2+} ion may be substituted for iron over the complete range of solid solutions, and these substitutions suppress the disproportionation reaction. In the system $Mg_yFe_{1-x-y}O$, the activation energy in the conductivity depends only on the O/Fe ratio, which suggests that the clusters remain and are nearly the same as in pure wüstite; the Mg^{2+} ions presumably tend to be segregated to the pure FeO pockets removed from the clusters. The similarity of the clusters in pure and Mg-doped wüstite has been demonstrated by Gourdin and Kingery[251].

(iii) Substitution of Na for Fe in wüstite creates Na^+-Fe^{3+} dipoles and an expansion of the lattice. Therefore the activation energy for conduction increases significantly with Na^+-ion concentration, not just with the O/Fe ratio[252–254].

(iv) All mixed-valent compounds containing iron and a counter cation in octahedral sites of a close-packed-cubic oxide-ion array, whether disordered in a solid-solution range or ordered into a line phase, give electrical conductivities that are described by a diffusional (hopping) model.

4.3 *The Series $CaFe_{2+n}O_{4+n}$*

The Fe–Ca–O system is of obvious interest to the steel industry. Reduction of the three calcium iron(III) oxides $2\,CaO \cdot Fe_2O_3$, $CaO \cdot Fe_2O_3$ and $CaO \cdot 2\,Fe_2O_3$ gives rise to several mixed-valence ternary compounds belonging to the $CaFe_2O_4$-$Fe_{1-x}O$ tie line[255–260].

From single crystals, prepared under precisely controlled temperature and oxygen pressure, Evrard et al.[259, 260] have shown that they form a series $CaFe_{2+n}O_{4+n}$ that can be classified by whether n is an integer, a half-integer, or n = 1/4. Those having an integral or half-integral n are intergrowth strucutres; they are orthorhombic (space group Cmcm) with $a \simeq 3.04$ Å, $b \simeq 10.0$ Å, and $c \simeq 10.0 + 2.65\,n$ Å for n = 1, 2, 3 and monoclinic for n = 3/2 and 5/2. The stacking sequences along c are:

n = 1, $CaFe_3O_5$: CWC*W*
n = 2, $CaFe_4O_6$: CWWC*W*W*
n = 3, $CaFe_5O_7$: CWWWC*W*W*W*
n = 3/2, $Ca_2Fe_7O_{11}$: W*CWC*W*
n = 5/2, $Ca_2Fe_9O_{13}$: W*CWWWC*W*

where C is a $CaFe_2O_4$ block, W is a wüstite (FeO) block, and an asterisk represents 180° rotation of a block about c. The compound $Ca_4Fe_9O_{17}$ (n = 1/4) has a completely different structure; it is monoclinic (space group C2) with four, five and six coordination for the iron.

Preliminary studies indicate that fast electron transfer is possible in $CaFe_3O_5$ between Fe^{2+} of the W blocks and Fe^{3+} of the C block (such assignment of valencies being valid below room-temperature).

An aside: It has been reported that Ca^{2+} ions substitute on tetrahedral sites of the spinel Fe_3O_4 to give $Ca_xFe_{1-x}[Fe_2]O_4$ over the compositional range $0 < x < 0.19$[261]. Chemically, such a substitution would be surprising, and the B-site isomer shift falls off more rapidly with x than the simple stoichiometric formula suggests. It is probable, therefore, that the Ca^{2+} ions enter the spinel structure on interstitial octahedral sites 16c together with two neighboring A-site vacancies to give

$$\{Ca_y^{2+}\}_{16c}\square_{2y}Fe_{1-2y}^{3+}[Fe_{1+4y}^{3+}Fe_{1-4y-z}^{2+}Ca_z^{2+}]O_4, \; x = y + z$$

5 Perovskite-Related Iron Oxides

5.1 Introduction

In the previous sections, interest was focussed on oxide systems containing Fe^{2+} and Fe^{3+} ions. In all these systems the iron ions were in a high-spin configuration with *localized* majority-spin electrons. The minority-spin electrons, formally associated with Fe^{2+} ions, occupied a narrow $Fe^{3+/2+}$: $3d^6$ band. Octahedral-site iron ions sharing common edges had $Fe^{3+/2+}$: $3d^6$ bands formed from orbitals of t_2 symmetry, and the bandwidth $w \simeq 2zb$ varied sensitively with the Fe-Fe separation and the number of nearest-neighbor iron atoms; it was also reduced by perturbations from impurity atoms or native defects that made the neighboring iron-atom potentials energetically inequivalent.

In the stoichiometric, cubic perovskites, the iron occupy only octahedral sites and couple to one another only through Fe–O–Fe interactions (see Fig. 17). The same is true of the tetragonal perovskite-rocksalt intergrowth structures such as La_2FeO_4. The cubic perovskites allow exploration of higher valence states; the intergrowth structures allow exploration of the full range of valence states. Unfortunately, systematic studies of the intergrowth iron oxides are not available. Moreover, although the Fe^{5+} valence state has been stabilized in the ordered perovskite La_2LiFeO_6[262], no examples of mixed $Fe^{4+/5+}$ valencies have been reported.

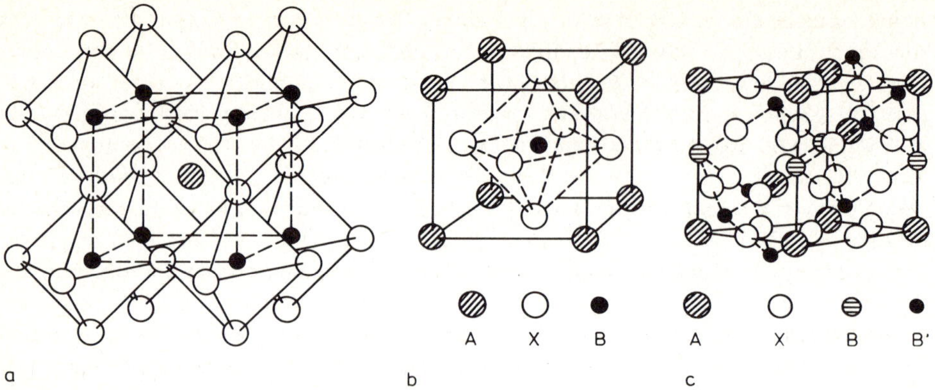

Fig. 17 a–c. The ideal cubic-perovskite structure for ABX_3 and ordered $A_2BB'X_6$

5.2 The System $Sr_{1-x}Ca_xFeO_3$

The Fe^{4+} ion has four d electrons, as do the Mn^{3+} and Ru^{4+} ions, so the perovskites $SrFeO_3$ and $CaFeO_3$ can be profitably compared with $LaMnO_3$ and $SrRuO_3$.

In semiconducting $LaMnO_3$, the Mn^{3+} ions have a localized high-spin configuration, and the orbitally twofold-degenerate ground state ^5E_g gives rise to a cooperative Jahn-Teller distortion responsible, at lower temperatures, for a magnetic ordering into ferromagnetic (001) planes coupled antiparallel to one another[263, 264)]. In metallic $SrRuO_3$, on the other hand, the Ru^{4+} ions have a low-spin configuration, and the compound exhibits an itinerant-electron ferromagnetism[264, 265)]. Since both the high-spin versus low-spin and the localized versus itinerant character of the d electrons depend on the degree of M–O covalent mixing, it is interesting to know whether the Fe^{4+} ions, with a covalent mixing intermediate to that of Mn^{3+} and Ru^{4+} ions, will have a high-spin or a low-spin configuration and whether its 3d electrons exhibit localized or itinerant properties.

Magnetic data provide direct information on the spin state of the Fe^{4+} ions. $CaFeO_3$ has a paramagnetic susceptibility typical of an antiferromagnet with a $\mu_{eff} = 5.6\ \mu_B$, somewhat larger than the spin-only value 4.9 μ_B for a high-spin d^4 configuration. Its Néel temperature is $T_N = 115$ K[266)]. The susceptibility of $SrFeO_3$, on the other hand, does not obey a Curie-Weiss law; and below a $T_N = 134$ K the iron moments order in a ferromagnetic helix propagating along the [111] axis with a fixed wave vector $|q| = 0.112 \cdot (2\pi/a_0)$[11, 267)]. The magnitude of the ordered Fe^{4+}-ion moment was reported to be $2.7 \pm 0.4\ \mu_B$ at 77 K in stoichiometric $SrFeO_3$; in $SrFeO_{2.9}$, the average moment at 4 K is only 2.3 μ_B, and subtraction of the Fe^{3+}-ion contribution (high-spin Fe^{3+}-ion Mössbauer signals were clearly evident) reduces the measured Fe^{4+}-ion moment to $1.6 \pm 0.3\ \mu_B$[268)]. These data demonstrate low-spin Fe^{4+} and high-spin Fe^{3+} configurations in oxygen-deficient $SrFeO_{3-x}$, and they imply an intermediate-spin configuration in stoichiometric $SrFeO_3$. On the other hand, Takeda et al.[269)] have extrapolated the high-field magnetization of the ferromagnetic system $SrFe_{1-x}Co_xO_3$ $(0.2 \leqslant x \leqslant 1)$ back to $SrFeO_3$ to obtain $\mu_{Fe} = 3.73\ \mu_B$ and a Mössbauer center shift that changes little for all x. They therefore argue for high-spin Fe^{4+} in $SrFeO_3$, which should then have a $\mu_{Fe} = 4\ \mu_B$. It would appear that the intermediate-spin state approaches the high-spin state in $SrFeO_3$. An intermediate-spin configuration has only been observed with itinerant-electron systems; localized-electron systems show classical spin-crossover behavior. Itinerant electrons in σ-antibonding (σ*) orbitals would be consistent with the fact that there is no cooperative Jahn-Teller distortion in either iron compound, such as is observed in $LaMnO_3$, and with the observation of metallic conductivity: a $\varrho = 10^{-3}$ Ωcm at room temperature is reported for $SrFeO_3$[270].

The change from high-spin Fe^{4+} in $CaFeO_3$ to intermediate-spin Fe^{4+} in stoichiometric $SrFeO_3$ and low-spin Fe^{4+} in $SrFeO_{3-x}$ may be understood as follows. First, an overview of the properties of transition-metal oxides with the perovskite structure[264] shows that high-spin configurations tend to have localized e electrons in the σ-antibonding orbitals; the change to itinerant character appears to just precede stabilization of low-spin configurations as the σ-bond covalent mixing is increased. This is not surprising as an increase in the covalent contribution to the M–O bond favors both electron delocalization and the low-spin configuration. However, the crystal-field splitting Δ_c responsible for the low-spin condition $\Delta_c > \Delta_{ex}$, where Δ_{ex} is the intraatomic exchange splitting, derives primarily from the difference in covalent bonding between e and t_2 orbitals. Where this difference is larger, the low-spin configuration is more probable. In $CaFeO_3$, a more acidic Ca^{2+} ion competes more strongly than the Sr^{2+} ion in $SrFeO_3$ for the O^{2-}: 2p electrons to reduce the Fe–O π bonding. Therefore we may unambiguously predict a greater stabilization of high-spin versus low-spin Fe^{4+} in $CaFeO_3$ than in $SrFeO_3$. Moreover, in $SrFeO_{3-x}$ the oxygen vacancies attract electrons to convert their two neighboring iron atoms from Fe^{4+} to Fe^{3+}; these Fe^{3+} ions provide the distinguishable Mössbauer signals. The O^{2-} ions on the opposite side of an Fe^{3+} ion no longer see equivalent iron near neighbors, so they bond more strongly to the ion of greater formal charge. This stronger covalent bonding, together with a perturbation that narrows the itinerant-electron σ bond, favors stabilization of a low-spin state over an intermediate-spin state. Thus we can understand how the introduction of oxygen vacancies can push an intermediate-spin state into a disproportionation into high-spin Fe^{3+} ions trapped at the vacancy and low-spin Fe^{4+} ions in the bulk. We conclude, therefore, that the evidence for a high-spin, itinerant-electron Fe^{4+}-ion configuration in $CaFeO_3$ is compatible with an intermediate-spin, itinerant-electron Fe^{4+}-ion configuration in $SrFeO_3$.

Consistent with this conclusion is the observation of an unusual feature in the Mössbauer spectrum of $CaFeO_3$, $T_N = 115$ K, on cooling below $T_t = 290$ K; the single Fe^{4+}-ion peak at $T > T_t$ is split into two spectra below T_t[271]. The tetragonal distortion of $CaFeO_3$ is too small to introduce an important quadrupole broadening. From the changes in center shifts (CS) and hyperfine fields H_i, see Table 3, it is possible to identify the transition as a disproportionation into ions of different charge; it is not a disproportionation into high-spin and low-spin configurations. Either reaction would be induced by a cooperative motion of the anions toward one iron subarray and away from the other as illustrated in Fig. 18. The high-spin versus low-spin disproportionation reaction was first proposed for $LaCoO_3$, which has a half-filled σ* band in the high-spin state, but it only appears in a temperature interval above room temperature in this compound[272]. In $CaFeO_3$, the disproportionation reaction is of the type

$$2\,Fe^{4+} \rightarrow Fe^{3+} + Fe^{5+}$$

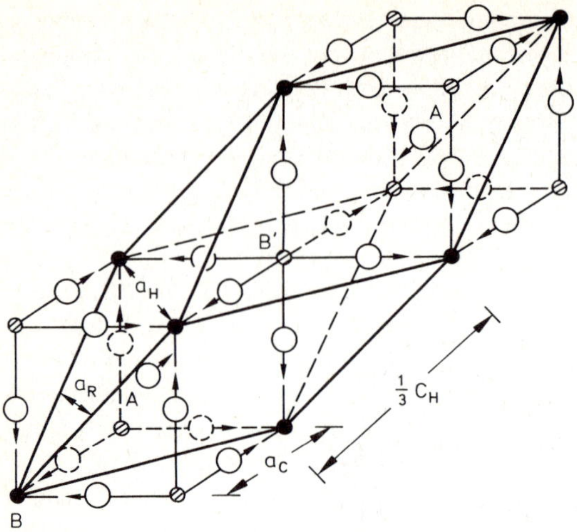

Fig. 18. Elementary unit cell and atomic displacements that induce disproportionation of an ABO_3 cubic perovskite into high-spin (B') and low-spin (B) or low-valence (B') and high-valence (B) cations to give rhombohedral R$\bar{3}$m symmetry

similar to that found below the melting point in $Pd^{2+}Pd^{4+}F_6$[273]. However, it is also clear that the disproportionation does not proceed, at least initially, as a full charge transfer per iron pair from one subarray to the other, so it is more correct to write the reaction as:

$$2\,Fe^{4+} \rightarrow Fe^{(4-\delta)+} + Fe^{(4+\delta)+}$$

the amount of charge transfer increasing with the magnitude of the cooperative anion displacement. Such a charge transfer represents a charge-density wave (CDW); its for-

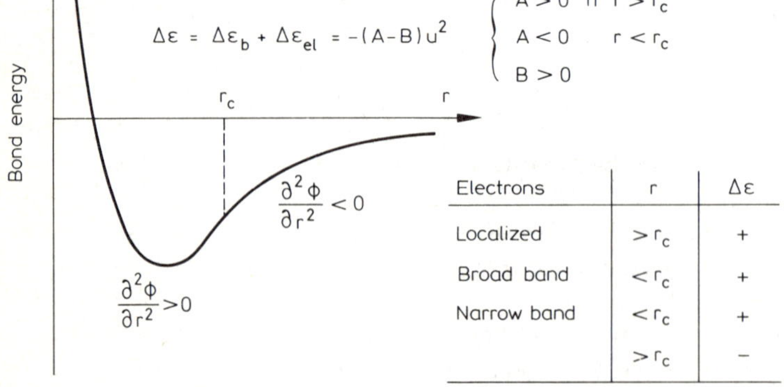

Fig. 19. The d-orbital contribution ϕ to the M–O interaction energy versus the M–O separation r

mation is a sure diagnostic of itinerant electrons (of e parentage) just as the Jahn-Teller distortion of $LaMnO_3$ is a diagnostic of localized e electrons at high-spin Mn^{3+} ions.

Figure 19 illustrates how a disporportionation reaction can be stabilized. The M–O interaction energy ϕ as a function of the atomic separation r for the cation-d-orbital contribution has the conventional Morse form with a minimum at r_m and an inflection at r_c. The M–O equilibrium separation $r_0 > r_m$ is determined by the sum of all the contributions to the electronic potential, which come primarily from s and p contributions of larger radial extension. A displacement of anions towards one set of cations and away from the others gives a change in energy

$$\Delta\varepsilon = \Delta\varepsilon_b + \Delta\varepsilon_{el} = -(A-B)u^2 \qquad (38)$$

where $\Delta\varepsilon_b$ comes from the d-electron contribution to the change in binding energy and $\Delta\varepsilon_{el}$ is an elastic-energy contribution coming from all other sources; u is the anion displacement normalized to half the cation-cation separation. The coefficient A is proportional to the curvature $\partial^2\phi/\partial r^2$, which changes sign at $r = r_c$. For $r_0 < r_c$, an $A < 0$ stiffens the elastic restoring force. However, for $r_0 > r_c$ an $A > 0$ softens the elastic restoring force and, where an $A > B$ exists, leads to a static disproportionation transition. Localized 3 d electrons have an $r_0 \gg r_c$, where A is small, and a large correlation energy U, which makes B large for the creation of polar states. Therefore an $A > B$ is not found with localized 3 d electrons. A maximum value of A occurs for a small, positive $(r_0 - r_c)$ where the bands are narrow. If U is large, a bandwidth $w \simeq 2zb > U$ may only be found where $r_0 < r_c$, so charge-transfer disproportionation reactions are rarely found with first-row transition-metal ions. However, a high-spin $3d^4$ configuration at octahedral sites contains an e-orbital degeneracy, so the effective charge-transfer energy U′ =

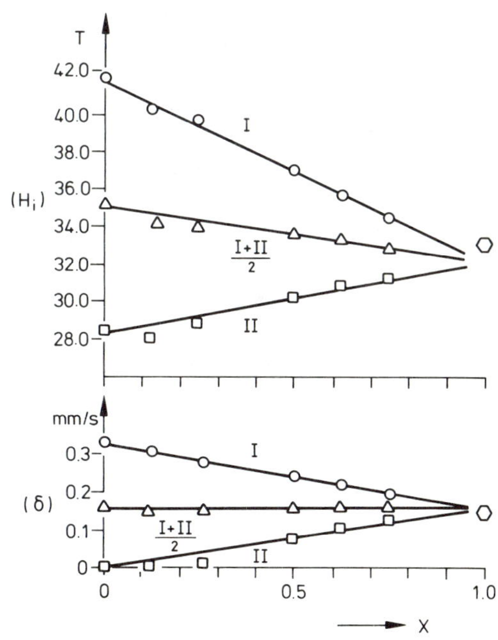

Fig. 20. Composition dependence of the two (*I* and *II*) hyperfine fields H_i and center shifts (*CS*) at 4 K in the system $Ca_{1-x}Sr_xFeO_3$, after[274)]

$U - \Delta_{ex}$ is reduced significantly by the intraatomic-exchange stabilization Δ_{ex}. Where $w \simeq U'$, this energy is reduced still further by screening, and a charge-transfer disproportionation reaction becomes possible. This most favorable situation applies in the case of $CaFeO_3$.

In the system $Ca_{1-x}Sr_xFeO_3$, the CS splitting and the transition temperature T_t decreases with increasing x, but the transition persists to $x > 0.75$ (see Fig. 20)[274]. This fact argues for a persistence of the high-spin Fe^{4+} state to large values of x and for a small $r_0 - r_c$ in $CaFeO_3$ so that A decreases with increasing σ^* bandwidth as x increases.

5.3 The System $Sr_{1-y}La_yFeO_3$

Substitution of La^{3+} for Sr^{2+} ions in $SrFeO_3$ does at least two things:

(i) it introduces a mixed $Fe^{4+/3+}$ valency and
(ii) it reduces the Fe–O covalent mixing because the La^{3+} ions compete more strongly for the O^{2-}: $2p^6$ electrons.

A reduced Fe–O covalent mixing would favor return to a high-spin Fe^{4+}-ion configuration, as in $CaFeO_3$, and Mössbauer data show disproportionation below a T_t in the range $0.1 < y < 0.6$[275]. A plot, Fig. 21(a), of the hyperfine field versus center shift for the two systems $Ca_{1-x}Sr_xFeO_3$ and $Sr_{1-y}La_yFeO_3$ at T = 4 K shows that both components of the disproportionation reaction fall on a single smooth curve; the single line of $SrFeO_3$ also falls on the line and at the center of the extrapolation, Fig. 21(b), for a virtual Fe^{4+} state, which indicates that any deviation from a high-spin to an intermediate-spin state in $SrFeO_3$ has surprisingly little influence on this relationship.

At room temperature, the Mössbauer spectrum of $Sr_{1-y}La_yFeO_3$ consists of only a single peak, somewhat broadened, with a center shift that increases linearly with y. At

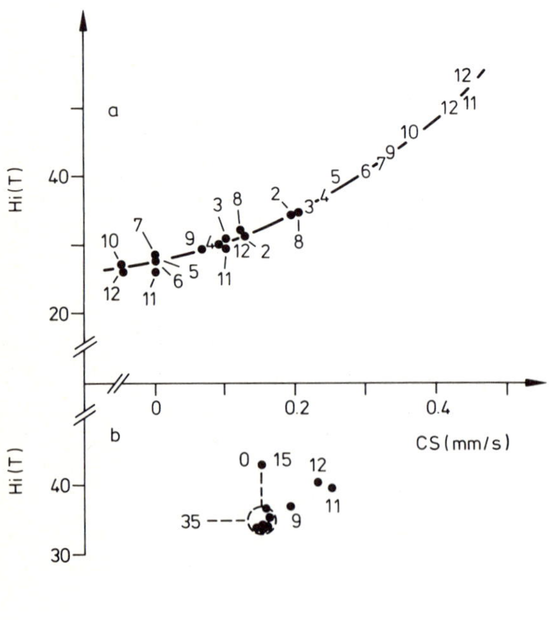

Fig. 21a, b. Mössbauer hyperfine field versus center shift data for the systems $Ca_{1-x}Sr_xFeO_3$ and $Sr_{1-y}La_yFeO_3$, **(a)** T = 4 K, **(b)** extrapolation to a virtual Fe^{4+}, i.e. $(I + II)/2$, after[275]

this temperature, fast electron transfer produces an average isomer shift for all the iron atoms. At lower temperatures, the La^{3+} ion would tend to capture an extra electron at its eight nearest neighbors, and below T_N the splitting of the spectra due to disproportionation gives intensities that are compatible with

$$0.5(1 + y)Fe^{(4-\delta)+} + 0.5(1 - y)Fe^{(4+\delta)+}$$

rather than with

$$0.5\ Fe^{(4-y-\delta)+} + 0.5\ Fe^{(4-y+\delta)+}$$

5.4 Perovskite-Related Structures

Perovskite-related structures include those containing ordered oxygen vacancies (brownmillerite, $Ca_2Fe_2O_5$), hexagonal polytypes ($6H - BaFeO_3$), and tetragonal perovskite-rocksalt intergrowths ($SrO \cdot 2\,SrFeO_3 = Sr_3Fe_2O_7$).

In the brownmillerite structure, the oxygen vacancies order in such a way that half the iron of $CaFeO_{2.5}$ are octahedrally coordinated and half are tetrahedrally coordinated. This ordering is an expression of the tetrahedral-site stability of Fe^{3+} ions. In a system like $La_{1-y}Ca_yFeO_{3-x}$, $x < y/2$, there is a tendency to form intergrowths of perovskite and brownmillerite structures; the oxygen vacancies do not remain randomly distributed. Mössbauer spectra at 4.2 K for this complex system exhibit three sextets[276)]; they have been interpreted in terms of 2 Fe^{3+} and one Fe^{5+}, or of an Fe^{3+} and the disproportiona-

Table 3. Mössbauer data[271)]

Oxide	T (K)	CS (mm/s)			Hi (T)		
$SrFeO_3$	298	0.054					
	4	0.146			33.1		
$CaFeO_3$	300	0.073					
		I	II	$\frac{I + II}{2}$	I	II	$\frac{I + II}{2}$
	4	0.34	0.00	0.17	41.6	27.9	34.8
$SrFeO_{2.86}$	298	(0.23)	0.021				
	4	0.303	–0.037	0.133	42.1	29.1	35.6
$Sr_3Fe_2O_{6.9}$	298	(0.277)	–0.033				
	4	0.317	–0.043	0.137	42.1	27.9	35.0
$BaFeO_{2.95}$	298	(0.475)	–0.065				
	4	0.635	–0.125	0.255	50.1	25.5	37.8

tion pair $Fe^{(4-\delta)+}$ and $Fe^{(4+\delta)+}$. Moreover, it appears that hopping terms remain slow ($\tau_h > 10^{-8}$ s) as the temperature is increased to about room temperature.

Intergrowths of rocksalt and perovskite are also known; they are represented by AO-$nABO_3$ with n = 1, corresponding to the K_2NiF_4 structure, being the most common. From Table 3, it appears that the $SrFeO_3$ perovskite layers in $Sr_3Fe_2O_{6.9}$ (n = 2) are sufficiently perturbed by the rocksalt layers that a high-spin Fe^{4+} configuration and disproportionation is stabilized.

Where the A cation is too large to be accomodated in the (BO_3) cage of the ABO_3 "cubic" perovskite structure, hexagonal polytypes are formed[264, 277]. These polytypes consist of different stackings of the close-packed AO_3 layers; in "cubic" perovskites this stacking is all cubic and in $CsNiCl_3$ it is all hexagonal. Cubic stacking leaves the all-oxygen octahedra sharing common corners; in hexagonal stacking they share one-dimensional chains of shared faces. In $CsNiCl_3$ the chains are not coupled, which allows the structure to accommodate the large Cs^+ ion. The transition from all-cubic to all-hexagonal stacking goes in stages, which allows for a variety of polytypes. The first stage from all-cubic to all-hexagonal stacking is the 6H–ABO_3 phase of hexagonal $BaTiO_3$; it consists of the stacking sequence hcchcc as illustrated in Fig. 22. For small x, $BaFeO_{3-x}$ crystallizes with this structure[278–280]. The phase diagram depends not only on the temperature and oxygen partial pressure[280–286], see Fig. 23, but also on traces of water vapor that may stabilize a tetragonally distorted all-cubic stacking for $0.38 < x < 0.47$ at 900 °C. At 900 °C, the 6 H polytype has an upper x limit of 0.32 for oxygen pressures of 100 bars; at 700 °C another polytype, Fig. 24, has been found[287–290]. In all these polytypes, the mixed valences are described by slow hopping ($\tau_h > 10^{-8}$ s)[287, 288, 290–292]. The magnetic properties have also been investigated for the 6 H polytype[293–295] and the 9 L polytype of Fig. 23.

The system $Ba_{1-x}Sr_xFeO_{3-y}$ has also been extensively studied[296, 297]. The structural, electrical and magnetic properties show a progressive evolution with x from those for $BaFeO_{3-y}$ to those for $SrFeO_{3-y}$.

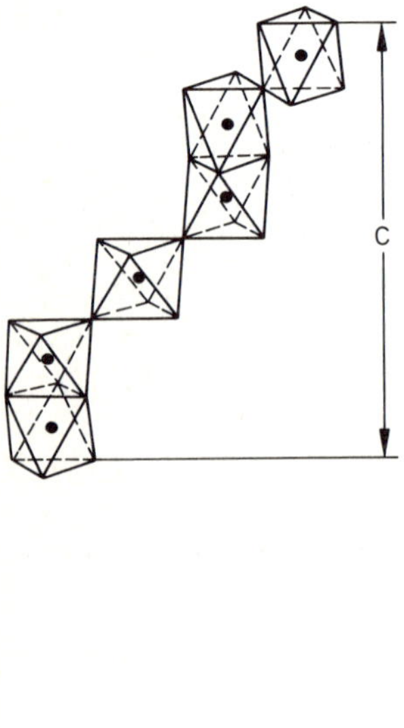

Fig. 22. The octahedral-site stacking sequence in the 6 H perovskite polytype of hexagonal $BaTiO_3$

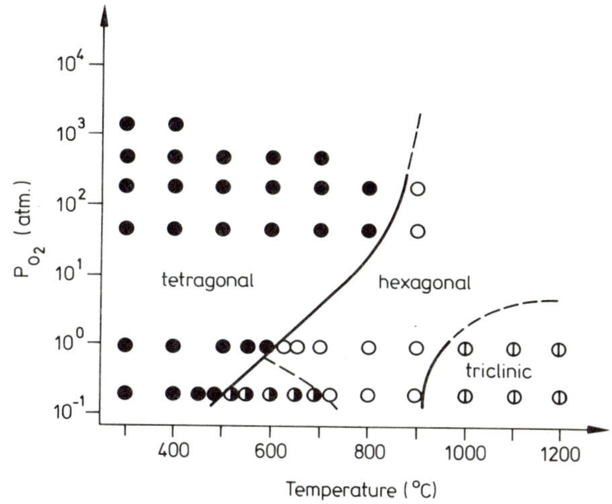

Fig. 23. Formation diagram of the products obtained by heating $BaFeO_3$ at various temperatures and oxygen partial pressures, after[286)]

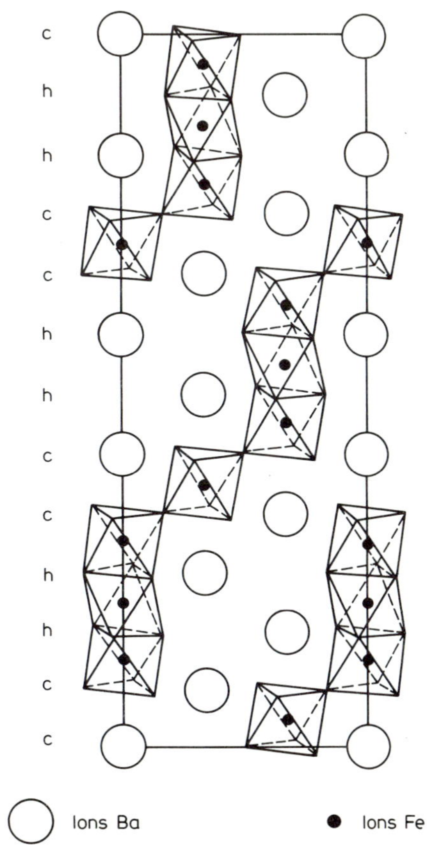

Fig. 24. The stacking sequence of the 12 L perovskite polytype

6 Miscellaneous Iron Oxides

A number of other oxides containing iron mixed-valence states are known; attention is drawn to some of these.

6.1 The Rare-Earth Ferrites $Ln_{1+n}Fe_{2+n}O_{4+3n}$

Whereas the larger rare-earth ions form the perovskites $LnFeO_3$, the smaller ions (Ln = Ho–Lu, Y) stabilize a series of structures $Ln_{1+n}Fe_{2+n}O_{4+3n}$ containing the Ln^{3+} ion in an octahedral site and the iron in trigonal bipyramidal sites. The end member, $LnFe_2O_4$, contains rare-earth basal planes alternating with a double-layer of edge-shared iron sites, see Fig. 25[298–300]. The other members of the series incorporate intergrowths of a similar rare-earth layer plus a single layer of Fe^{3+} ions in trigonal-bipyramidal sites, Fig. 26. If the rare-earth, single-iron, and double-iron layers are designated

$$U = LnO_{1.5} \quad V = FeO_{1.5} \quad W = Fe_2O_{2.5}$$

then the stacking sequences are as follows[300]:

$LnFe_2O_4$:	UWUWUW	space group	$R\bar{3}m$
$Ln_2Fe_3O_7$:	UVUWUVUW		$P6_3/mmc$
$Ln_3Fe_4O_{10}$:	UVUVUWUVUVUWUVUVUW		$R\bar{3}m$
$Ln_4Fe_5O_{13}$:	UVUVUVUWUVUVUVUW		$P6_3/mmc$

In these structures, the double layers of iron contain the mixed-valence ratio $Fe^{2+}/Fe^{3+} = 1$, and the Fe–Fe distance of about 3.1 Å is short enough for important Fe–Fe interactions; but it is larger than in magnetite (2.97 Å) and there are only three instead of six iron nearest neighbors. It is therefore not surprising to find a large anisotropy in the conductivity (at room temperature $\varrho_\perp \simeq 10\,\varrho_\parallel$) with the basal-plane conductivity $\varrho_\perp \simeq 10\ \Omega cm$ some three orders of magnitude larger than is found in magnetite[301]. The compounds tend to be oxygen-deficient[302–305], and electron trapping at oxygen vacancies would contribute a perturbation of the $Fe^{3+/2+}$: $3d^6$ band to help stabilize a small-polaron, slow-hopping regime. All compounds exhibit an activation energy in the conductivity.

Phase diagrams[302–307] and magnetic properties[301, 308–313] have also been investigated. Magnetic coupling between the iron double layers is weak, but within the layers it is strong.

Mössbauer data[308, 314–316] indicate slow electron hopping below room temperature with some evidence for an ordering of Fe^{2+} and Fe^{3+} ions below 220 K in YFe_2O_4, but fast hopping ($\tau_h < 10^{-8}$ s) at higher temperatures.

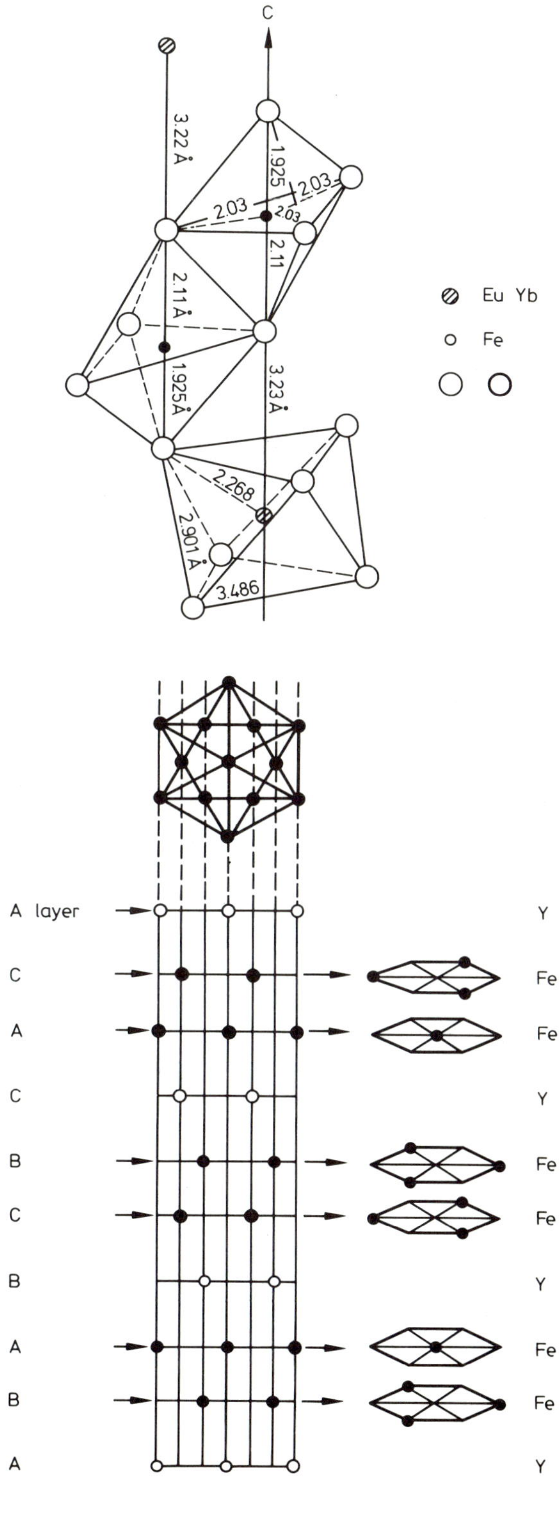

Fig. 25. Atomic stacking of $Yb_{0,5}Eu_{0,5}Fe_2O_4$, after[300)]

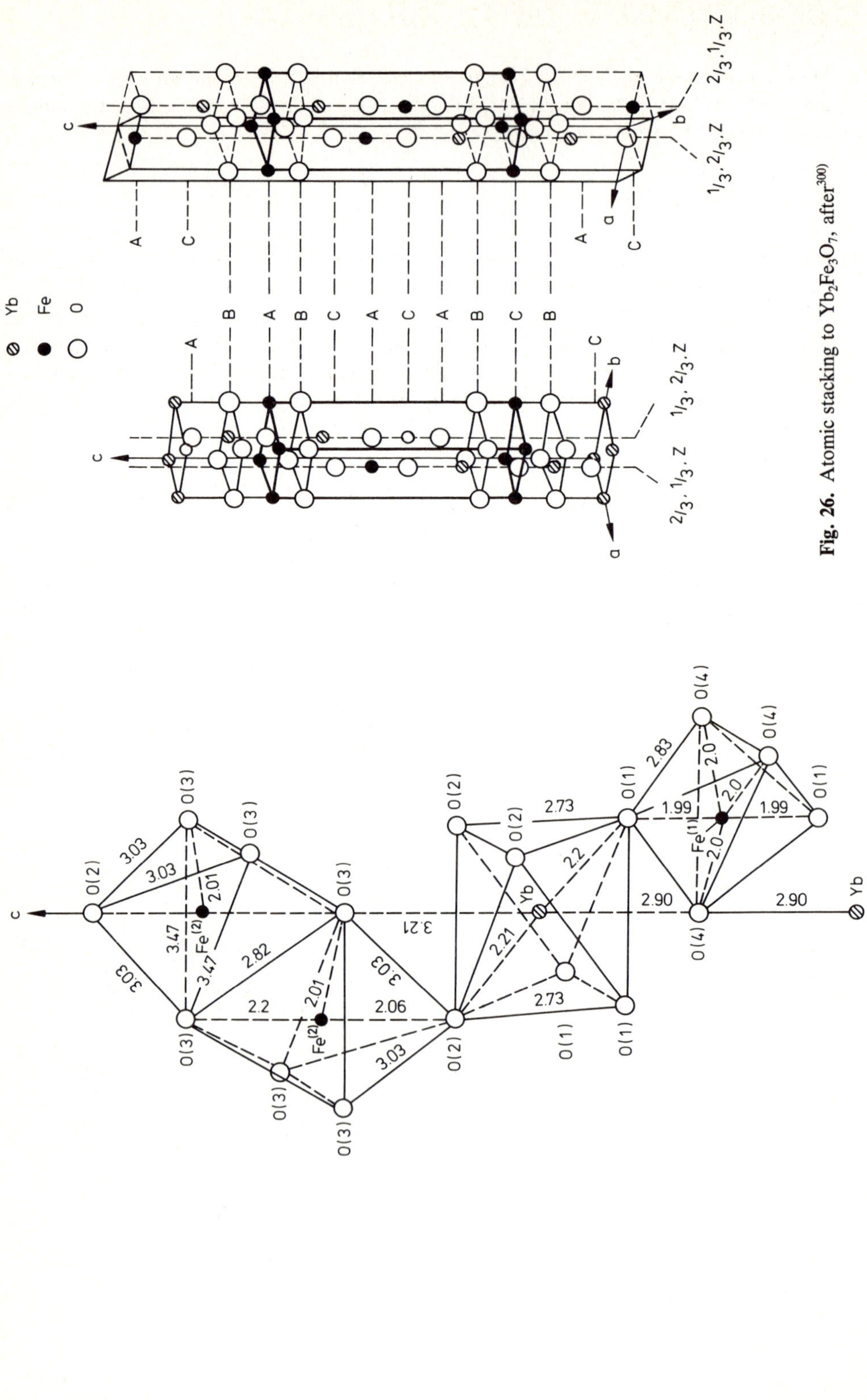

Fig. 26. Atomic stacking to $Yb_2Fe_3O_7$, after[300]

6.2 *Iron Phosphates and Oxyphosphates*

Phosphorus-containing iron ores are sometimes used in the steel industry, so an investigation of the Fe–P–O system at 800–1000 °C has been made. Some of the new compounds discovered in this work contain mixed iron valencies.

6.2.1 $Fe_2(PO_4)O$[317]

Prepared at 900 °C, the orthorhombic unit cell contains Fe^{2+} and Fe^{3+} ordered into inequivalent, distorted octahedra, see Fig. 27. Although the Fe^{3+} and Fe^{2+} octahedra share common faces with an Fe^{3+}–Fe^{2+} separation of only 2.92 Å, nevertheless Mössbauer measurements up to 600 °C indicate distinguishable Fe^{2+} and Fe^{3+} ions with only a slow hopping ($\tau_h > 10^{-8}$ s) between them[318]. This comes not only from the fact that the two sites are energetically inequivalent, but also from an orientation of the t_{2g} orbitals that is less favourable to overlaps than for edge-sharing octahedra.

6.2.2 $Fe_9(PO_4)O_8$ and a Related Compound[319]

This compound, recently synthesized[319], presents a block structure with an alternate stacking of FeO and Fe_3PO_6 slabs. Unexpectedly, within the Fe_3PO_6 slabs, where there

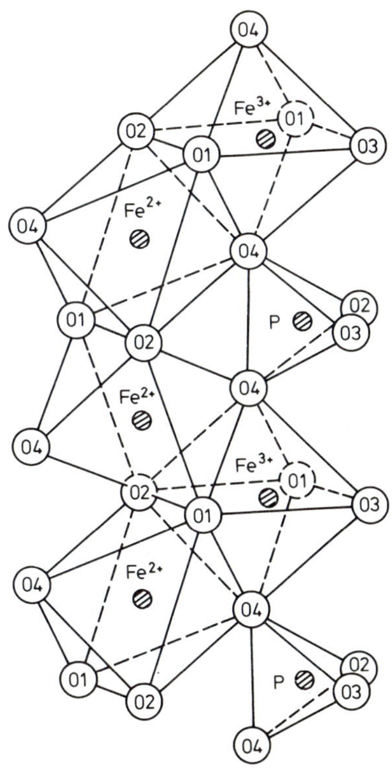

Fig. 27. Face-shared arrangement of Fe^{2+} and Fe^{3+} octahedra in the Fe_2PO_5 structure, after[317]

are rows of alternating FeO_4 and PO_4 tetrahedra, iron is divalent (according to the Fe–O distances provided by the refinement characterized by a reliability factor R = 0.033). Consequently, the electron holes are inside the FeO blocks and, indeed, a fast electron transfer is recorded on the Mössbauer spectra above 100 °C. $Fe_9(PO_4)O_8$ is antiferromagnetic with $T_N = 217$ K; unlike wüstite, it does not disproportionate on heating.

A somewhat analogous stacking is observed in $Fe_7(SiO_4)O_6$, first reported by Smuts et al.[320)] to have 2 wüstite blocks $(FeO)_4$ and $(FeO)_8$ alternating with 2 slabs of composition $FeSiO_4$. According to more recent results[321)], a high disorder is observed in the $FeSiO_4$ blocks as shown by both Weissenberg patterns and electron microscopy.

6.3 *Cronstedtite*

The block iron silicate $|Si_{2-x}Fe_x^{3+}|\{Fe_{3-x}^{2+}Fe_x^{5+}\}O_5(OH)_4$, cronstedtite, is representative of several mixed-valent minerals. Its structure consists of alternate layers of corner-shared tetrahedra and edge-shared octahedra. The tetrahedra Si^{4+} and Fe^{3+} ions are coordinated by four oxide ions, the octahedral Fe^{2+} and Fe^{3+} ions share three common oxide ions with tetrahedral-site layers on one side and on the other are coordinated by three hydroxyl ions that hydrogen bond to the next tetrahedral-site layer. Mössbauer data at 35 K are able to resolve distinguishable Fe^{2+} and Fe^{3+} ions on the octahedral sites; but in the range $300 < T < 500$ K a rapid ($\tau_h < 10^{-8}$ s) charge transfer was found[322)].

7 Conclusions

This review of some available data on mixed $Fe^{3+/2+}$ and $Fe^{4+/3+}$ valences in iron oxides leads us to the following conclusions:

(i) It is useful to distinguish three types of mobile electrons:
(a) *Small polarons:* mobile electrons "dressed" in a local deformation and having an activated mobility $u_h = (eD_0/kT)\exp(-\Delta H_m/kT)$
(b) *Itinerant electrons:* electrons belonging equally to all like atoms on energetically equivalent lattice sites and having a mobility $u_i = e\tau_s/m^*$
(c) *Intermediate electrons:* mobile electrons that move from iron to iron with a diffusional motion at high temperatures, as small polarons, but with a $\Delta H_m < kT$ so that $u_h \simeq (eD_0/kT)$; at low temperatures strong electron-phonon interactions create random small polarons trapped at impurities or defects if in small concentration or ordered (short and/or long-range) by electron-electron electrostatic correlation.

(ii) Both Fe^{2+} and Fe^{3+} are found in oxides in a high-spin state; the majority-spin electrons are localized, but the minority-spin electrons introduced by Fe^{2+} may – on an array of edge-shared, energetically equivalent octahedral sites – be small polarons ($Fe_{1-x}O$), itinerant electrons ($Ge[Fe_2]O_4$), or intermediate electrons (Fe_3O_4).

(iii) An Fe^{4+} ion in an oxide may have either a high-spin or a low-spin or an intermediate-spin state; and an array of Fe^{4+} ions in energetically equivalent, corner-shared octahedral sites having approximately 180° Fe–O–Fe bonding has itinerant σ-antibonding (σ*) electrons.

(iv) Whereas cooperative Jahn-Teller distortions are diagnostic of localized configurations at transition-metal ions in solids, a charge-density wave (CDW) at low temperatures is diagnostic of itinerant electrons. The CDW stabilized by an array of high-spin Fe^{4+} ions is a charge-transfer disproportionation toward Fe^{3+} and Fe^{5+}.

(v) Electron transfer between iron ions of different valence on sites having an energy inequivalence $\Delta = \Delta H - T\Delta S$ is modulated by the factor $\exp(-\Delta/kT)$, which may go to unity at higher temperatures.

(vi) Foreign-atom solutes and native defects tend to stabilize and to trap small polarons in oxides containing Fe^{3+} and Fe^{2+} ions.

(vii) Short-range (dynamic) electron-phonon interactions with electron-electron electrostatic correlations change the statistics for the Seebeck coefficient from Eq. (15) to Eq. (16); they also tend to stabilize regions having a valence ratio $Fe^{2+}/Fe^{3+} \simeq 1$. This latter phenomenon has led Mössbauer spectroscopists to postulate the formation of Fe^{3+}–Fe^{2+} pairs with more rapid electron hopping within than between pairs, but this configuration is not a unique explanation.

(viii) Interpretation of intermediate-electron processes depends critically on the time scale of the measurements. Low-temperature Fe_3O_4, for example, does not appear to distinguish iron of different valence state on the time scale of a neutron-diffraction experiment, but it does on the time scale of NMR and Mössbauer measurements. Confusion on this point has led to considerable controversy in the literature.

(ix) Fast electron transfer introduces a ferromagnetic double-exchange component to Fe–Fe interactions that disappears if the hopping time becomes slow ($\tau_h > 10^{-6}$ s) as a result of small-polaron formation.

(x) The Fe^{3+} ion is readily accomodated in a tetrahedral, trigonal bipyramidal, or octahedral coordination of oxide ions.

(xi) The simultaneous appearance of mixed valences on two different cations occurs where there are overlapping redox bands; such overlap is more probable where one of the redox bands corresponds to itinerant electrons in a band of finite width, as in $Fe[FeMo]O_4$. In the ferrospinels, such an overlap leads to important Fe_A–O–M_B–O–Fe_B charge-transfer interactions; the Fe_A–O–O–F_A interactions remain weak.

(xii) The bottom of the $Sn^{3+}: 5s^1$ band is close enough to the top of the $Fe_A^{2+}: 3d^6$ band in a spinel for strong admixing of the charge-transfer excited state $Fe_A^{3+} + Sn^{3+}$ into the ground state and a finite ferromagnetic Fe_A^{2+}–O–Sn_B–O–Fe_A^{2+} interaction.

Not considered in this review are the removal of energetic equivalence by an applied field or stress where, as in a spinel, the energetically equivalent sites have differently oriented crystal-field axes. This lifting of site degeneracy coupled with charge transfer between mixed-valence states leads to such phenomena as "magnetic after effect"[323] and "photoinduced anisotropy"[324].

8 References

1. Burns, R. G., Nolet, D. A., Parkin, K. M., McCammon, C. A., Schwartz, K. B.: in Mixed Valence Compounds (Brown, D., Ed.), Reidel Publ. 1980, p. 295
2. Blakemore, R. P.: Science *190,* 377 (1975)
3. Mizushima, K., Tanaka, M., Asai, A., Iida, S., Goodenough, J. B.: J. Phys. Chem. Sol. *40,* 1129 (1979)
4. Goodenough, J. B.: Metallic Oxides, in Progr. Solid State Chem. *5,* 145 (1971)
5. Robin, M. B., Day, P.: Adv. Inorg. Chem. Radiochem. *10,* 247 (1967)
6. Mayoh, B., Day, P.: J. Chem. Soc. Dalton Trans. 846 (1974)
7. Marezio, M., McWhan, D. B., Dernier, P. D., Remeika, J. P.: J. Solid State Chem. *6,* 213 (1973)
   The Verwey Transition, Phil. Mag. (Special Issue) *42B,* 325 (1980)
8. Shirane, G., Pickart, S. J., Nathans, R., Ishikawa, Y.: J. Phys. Chem. Solids *10,* 35 (1959)
9. Cox, D. E., Takei, W. J., Shirane, G.: ibid. *23,* 863 (1962)
10. Gupta, M. P., Kanetkar, S. M., Date, S. K., Nigakevar, A. S., Sinha, A. P. B.: J. Phys. *C12,* 2401 (1979)
11. Gallagher, P. K., McChesney, J. B., Buchanan, D. N. E.: J. Chem. Phys. *41,* 2429 (1964)
12. Berkooz, O., Malamud, M., Shtrikman, S.: Solid State Comm. *6,* 185 (1968)
13. Brown, D. B., Wrobleski, J. T.: in Mixed valence compounds (Brown, D., Ed.), Reidel Publ. 1980, p. 243
14. Greenwood, N. N., Gibb, T. C.: Mössbauer spectroscopy, Chapman & Hall, London 1971
15. Kaplan, T. A., Dwight, K., Lyons, D. H., Menyuk, N.: J. Appl. Phys. Suppl. *32,* 135S (1961)
16. Darken, L. S., Gurry, R. W.: J. Am. Chem. Soc. *16,* 1398 (1945)
17. Dieckmann, R., Schmalzried, H.: Ber. Bunsenges. Phys. Chem. *81,* 344 (1977)
18. Colombo, U., Gazzarini, F., Lanzavecchia, G.: Mat. Sci. Eng. *2,* 125 (1967)
19. Pritchard, A. M., Haddon, J. R., Walton, G. N.: Corrosion Sci. *11,* 11 (1971)
20. Garcia-Gonzales, M. L., Grange, P., Delmon, B.: 8th Int. Symp. Reactivity Solids, Göteborg 1976
21. Moukassi, M., Gerardin, R., Dupre, B., Gleitzer, C.: J. Mater. Sci. *17,* 1213 (1982)
22. Schrader, R., Büttner, D.: Z. Anorg. Allg. Chem. *320,* 205 (1963)
23. Verwey, E. J. W., De Boer, J. H.: Rec. Trav. Chem. *55,* 531 (1936)
24. Millar, R. W.: J. Am. Chem. Soc. *51,* 215 (1929)
25. Landolt-Bornstein, New Series, Vol. III/3b, Springer Verl. Berlin 1970
26. Okamura, T.: Sci. Rep. Tohoku Univ. *21,* 231 (1931)
27. Evans, B. J., Westrum, E. H.: Phys. Rev. *B5,* 3791 (1972);
    Mitzui, M., Tado, S., Chikazumi, S.: J. Phys. Soc. Japan *43,* 47 (1977)
    Rigo, M. O., Mareche, J. F., Brabers, V. A. M.: Phil. Mag. *B 48,* 421 (1983)
28. Miles, P. A., Westphal, W. B., von Hippel, A.: Rev. Mod. Phys. *29,* 289 (1957)
29. Verwey, E. J. W.: Nature *144,* 327 (1939);
    Verwey, E. J. W., Haaymann, P. W.: Physica *8,* 979 (1941);
    Verwey, E. J. W., Romeijn, R. C.: J. Chem. Phys. *15,* 18 (1947)
30. Bickford, L.: Phys. Rev. *78,* 449 (1950)
31. Néel, L.: Ann. Phys. *3,* 138 (1948)
32. Shull, C. G., Wollan, E. O., Koehler, W. C.: Phys. Rev. *84,* 912 (1951)
33. Srivastava, C. M., Shringi, S. N., Babu, M. V.: Phys. Stat. Sol. *65A,* 731 (1981)
34. Torrie, D. H.: Solid State Comm. *5,* 715 (1967)
35. Goodenough, J. B.: Magnetism and the Chemical Bond, Interscience & Wiley, New York 1963
36. Schlegal, A., Alvarado, S. F., Wachter, P.: J. Phys. (Paris) Colloque *40,* C12–1157 (1979)
37. Alvarado, S. F., Erbudak, M., Munz, P.: Phys. Rev. *B14,* 2740 (1976)
38. Cullen, J. R., Callen, E.: J. Phys. (Paris) Colloque *32,* C1–1110 (1971)
39. Samara, G. A., Giardini, A. A.: Phys. Rev. *186,* 577 (1969)
40. Goodenough, J. B.: ibid. *171,* 466 (1968)
41. Umemura, S., Iida, S.: J. Phys. Soc. Japan *40,* 679 (1976)
42. Siemons, W. J.: IBM J. Res. Dev. (May) 245 (1970)

43. Galeczki, G., Buckwald, R. A., Hirsch, A. A.: Solid State Comm. *23,* 201 (1977)
44. Boekema, C.: Phil. Mag. *B42,* 409 (1980)
45. Srinivasan, G., Srivastava, C. M.: Phys. Stat. Sol. *103b,* 665 (1981)
46. Srivastava, C. M.: Bull. Mater. Sci. *5,* 247 (1983)
47. Terukov, E. I., Reichelt, W., Ihle, D., Oppermann, H.: Phys. Stat. Sol. (b) *95,* 491 (1979)
48. Samara, G. D.: Phys. Rev. Lett. *21,* 795 (1968)
49. Kakudata, Y., Mori, N., Kino, Y.: J. Magn. Magn. Mater. *12,* 22 (1979)
50. Whall, T. E., Rigo, M. O., Jones, M. R. B., Pointon, A. J.: J. Phys. (Paris) Colloque *38,* C1–229 (1977)
51. Wu, C. C., Mason, T. O.: J. Am. Ceram. Soc. *64,* 520 (1981)
52. Sawatzki, G. A., Coey, J. M. D., Morrish, A. H.: J. Appl. Phys. *40,* 1402 (1969)
53. Van Diepen, A. M.: Physica *86–88,* 955 (1977)
54. Balko, B., Hoy, G. R.: J. Phys. (Paris) Colloque *37,* C6–89 (1976)
55. Graener, H., Rosenberg, M., Whall, T. E., Jones, M. R.: Phil. Mag. *B40,* 389 (1979)
56. Mott, N. F.: Festkörperprobleme *XIX,* 331 (1979)
57. Bertaut, E. F.: Bull. Soc. Franç. minéral. et crist. *79,* 276 (1956)
58. Goodenough, J. B.: J. Appl. Phys. Suppl. *33,* 1197 (1962)
59. Goodenough, J. B.: in Solid State Chemistry (ed. Rao, C. N. R.), M. Dekker publ., N.Y. 1974, p. 321
60. Tombs, N. C., Rooksby, H. P.: Acta Cryst. *4,* 474 (1951)
61. Smiltens, J.: J. Chem. Phys. *20,* 990 (1952)
62. Calhoun, B. A.: Phys. Rev. *94,* 1577 (1954)
63. Abrahams, S. C., Calhoun, B. A.: Acta Cryst. *6,* 105 (1953); *8,* 257 (1955)
64. Hamilton, W. C.: Phys. Rev. *110,* 1050 (1958)
65. Ono, K., Ishikawa, Y., Ito, A., Hirahara, E.: J. Phys. Soc. Japan *17,* Sup B1, 125 (1962)
66. Yamada, Y., Suzuki, K., Chikazumi, S.: Appl. Phys. Lett. *13,* 172 (1968)
67. Rado, G. T., Ferrari, J. M.: Phys. Rev. *B12,* 5166 (1975)
68. Yoshida, J., Iida, S.: J. Phys. Soc. Japan *42,* 230 (1977)
69. Iida, S., Mizushima, K., Mizoguchi, M., Umemura, S., Yoshida, Y.: J. Appl. Phys. *49,* 1455 (1978)
70. Yamada, Y.: Phil. Mag. *B42,* 373 (1980)
71. Daniels, J. M., Rosencwaig, A.: J. Phys. Chem. Solids *30,* 1561 (1969)
72. Ihle, D., Lorentz, B.: Phil. Mag. *B42,* 337 (1980)
73. Chakraverty, B. K.: ibid. *B42,* 473 (1980)
74. Rubinstein, M., Forester, D. W.: Solid State Comm. *9,* 1675 (1971); Rubistein, M., Strauss, G. H., Brumi, F. J.: AIP Conf. Proc. *10,* 1384 (1972)
75. Kovtun, N. M., Shamyakov, A. A.: Solid. State Comm. *13,* 1345 (1973)
76. Hargrove, R. S., Kündig, W.: ibid. *8,* 303 (1970)
77. Mada, J., Iida, S.: J. Phys. Soc. Japan *39,* 1627 (1975)
78. Umemura, S., Iida, S.: ibid. *47,* 458 (1979)
79. Mizoguchi, M.: ibid. *44,* 1501, 1512 (1978)
80. Iizumi, M., Koetzle, T. F., Shirane, G., Chikazumi, S., Matsui, M., Todo, S.: Acta Cryst. *B38,* 2121 (1982)
81. Iida, S.: Phil. Mag. *B42,* 349 (1980)
82. Cullen, J. R.: ibid. *B42,* 387 (1980)
83. Iida, S.: Private Communication
84. Goto, Y.: Japan J. Appl. Phys. *3,* 739 (1964)
85. Coey, J. M. D., Morrish, A. H., Sawatzky, G. A.: J. Phys. (Paris) Colloque *32,* C1–271 (1971)
86. Feitknecht, W., Gallagher, K. J.: Nature *228,* 548 (1970)
87. Gallagher, K. J., Feitknecht, W., Mannweiller, U.: ibid. *217,* 1118 (1968)
88. Colombo, U., Gazzarini, F., Lanzavecchia, G., Sironi, G.: ibid. *219,* 1036 (1968)
89. Topsoe, H., Dumesic, J. A., Boudart, M.: J. Phys. (Paris) Colloque *35,* C6–411 (1974)
90. Haneda, K., Kojima, H.: ibid. *40,* C2–583 (1979)
91. Landolt-Bornstein, New Series, Vol. *III/12b,* Springer Verl. Berlin 1980
92. Henry, W. E., Boehm, M. J.: Phys. Rev. *101,* 1253 (1956)
93. Ferguson, G. A., Hass, M.: ibid. *112,* 1130 (1958)
94. Takei, H., Chiba, S.: J. Phys. Soc. Japan *21,* 1255 (1966)

95. Annersten, H., Hafner, S. S.: Z. Kristallog. *137,* 321 (1973)
96. Gillot, B.: J. Phys. Chem. Solids *40,* 261 (1979)
97. Portier, J., Claverie, J., Dexpert, H., Olazcuaga, R., Hagenmuller, P.: C.R. Acad. Sci. *270C,* 2142 (1970)
98. Franke, H., Rosenberg, M., Whall, T. E., Jones, M. R. B.: J. Magn. Magn. Mater. *6,* 151 (1977)
99. Franke, H., Rosenberg, M., Whall, T. E., Jones, M. R. B.: ibid. *7,* 223 (1978)
100. Rigo, M. O., Courrier, R., Pointon, A., Foct, J.: Rev. Chimie Miné. *14,* 33 (1977)
101. Paul, A., Basu, S.: Trans. J. Brit. Ceram. Soc. *73,* 167 (1974)
102. Popov, G. P., Ilinova, G. N.: Zh. Fiz. Khim. *45,* 541 (1971)
103. Miyata, N.: J. Phys. Soc. Japan *16,* 1291 (1961)
104. Srivastava, C. M., Shringi, S. N., Srivastava, R. G., Nanadikar, N. G.: Phys. Rev. *14B,* 2032 (1976)
105. Claude, J. M.: Unpublished work
106. Fleet, M. E.: Acta Cryst. *B37,* 917 (1981)
107. Verwey, E. J. W., Heilmann, E. L.: J. Chem. Phys. *15,* 174 (1947)
108. Guillaud, C.: J. Phys. Radium *12,* 239 (1951)
109. Gorter, E. W.: Philips Res. Rept. *9,* 295 (1954)
110. Dobson, D. C., Linnet, J. W., Rahman, M. L.: J. Phys. Chem. Solids *31,* 2727 (1970)
111. Srivastava, C. M., Shringi, S. N., Srivastava, R. G.: Phys. Rev. *14B,* 2041 (1976)
112. Ok, H. N., Evans, B.: ibid. 2956
113. Dickof, P. A., Schurer, P. J., Morrish, A. H.: ibid. *22B,* 115 (1980)
114. Samkhralov, A. A., Ranstamov, A. G.: Soviet Phys. – Solid State *7,* 961 (1965)
115. Dehe, G., Suwalski, J., Wieser, E., Kabisch, R.: Phys. Stat. Sol. *65a,* 669 (1981)
116. Oles, A.: Acta Phys. Polon. *30,* 125 (1966)
117. Gamari-Seale, H., Papamantellos, P.: Phys. Stat. Sol. *29,* 323 (1968)
118. Alebouyeh, A.: Thesis Nancy (1981)
119. Kovtun, E. F., Cizova, Z. I.: Neorg. Mater. *14,* 379 (1978)
120. Braun, P. B.: Nature *170,* 1123 (1952)
121. Schieber, M.: J. Inorg. Nucl. Chem. *26,* 1363 (1964)
122. Belov, V. F., Kirichok, P. P., Glotov, V. G., Podvalnykh, G. S., Bondarev, D. E., Shipko, M. N.: Izv. Akad. Nauk SSSR, Neorg. Mater. *7,* 2106 (1971)
123. Simonet, W., Hermosin, A.: IEEE Mag. *14,* 903 (1978)
124. Dormann, J. L.: J. Phys. (Paris) Colloque *41,* C1–175 (1980)
125. Magnetic oxides, Ed. Craik, D. J., J. Wiley 1975
126. Miyata, N.: J. Phys. Soc. Japan *16,* 206 (1961)
127. Parker, R., Tinsley, C. J.: Proc. Int. Conf. Ferrites, Univ. Tokyo Press, Tokyo, 592 (1971)
128. Griffith, B. A., Elwell, D., Parker, R.: Phil. Mag. *22,* 163 (1970)
129. Nicolau, P., Bunget, I. N., Rosenberg, M., Belciu, I. N.: IBM J. Res. Dev. *14,* 248 (1970)
130. Linnett, J. W., Rahman, M.: J. Phys. Chem. Solids *33,* 1465 (1972)
131. Yearian, H. J., Kortright, J. M., Langenheim, R. H.: J. Chem. Phys. *22,* 1196 (1954)
132. Francombe, M. H.: J. Phys. Chem. Solids *3,* 37 (1957)
133. Franke, H., Rosenberg, M.: J. Magn. Magn. Materials *9,* 74 (1979)
134. Robbins, M., Wertheim, G. K., Cherwood, R. C., Buchanan, D. N. E.: J. Phys. Chem. Solids *32,* 717 (1971)
135. Goodenough, J. B.: ibid. *25,* 151 (1964)
136. Goodenough, J. B.: in New Developments in Semiconductors (Wallace, P. R., Harris, R., Zuckerman, J. J., eds.), Noordhoff, Leyden 1973, p. 107
137. Yafet, Y., Kittel, C.: Phys. Rev. *87,* 290 (1952)
138. Lyons, D., Kaplan, T. A.: ibid. *120,* 1580 (1960)
139. Prince, E.: Acta Cryst. *10,* 554 (1957)
140. Gaballah, I., Courtois, A., Jeannot, F., Gleitzer, C.: C.R. Acad. Sci. *280C,* 1367 (1975)
141. Chassagneux, F., Rousset, A.: J. Solid State Chem. *16,* 161 (1976)
142. Dehe, G., Seidel, B., Malzer, K., Michalk, C.: Phys. Stat. Sol. *31a,* 439 (1975)
143. Mason, T. O., Bowen, H. K.: J. Am. Ceram. Soc. *64,* 86 (1981)
144. Lotgering, F. K.: J. Phys. Chem. Solids *23,* 1153 (1962)
145. Slack, G. A.: Phys. Rev. *134A,* 1268 (1964)

146. Wakihara, M., Shimizu, Y., Katsura, T.: J. Solid State Chem. *3,* 478 (1971)
147. Bernier, J., Poix, P.: Ann. Chim. *2,* 1981 (1967)
148. Abe, M., Kawachi, M., Nomura, S.: J. Solid State Chem. *10,* 351 (1974)
149. Rogers, D. A., Arnott, R. J., Wold, A., Goodenough, J. B.: J. Phys. Chem. Solids *24,* 347 (1963)
150. Rossiter, M. J.: ibid. *26,* 775 (1965)
151. Reuter, B., Riedel, E.: Ber. Bunsenges. Phys. Chem. *72,* 880 (1968)
152. Lee, J. D., Schroeer, D.: J. Phys. Chem. Solids *37,* 739 (1976)
153. Smit, J., Wijn, H. P. J.: Ferrites: Physical Properties of Ferrimagnetic Oxides in Relation to their Technical Applications, John Wiley, N.Y. & Philips Tech. Libr., London 1959
154. Bernier, J., Poix, P.: Ann. Chim. *2,* 1981 (1967)
155. Basile, F., Djega-Mariadassou, C., Poix, P.: Mat. Res. Bull. *8,* 619 (1973)
156. Djega-Mariadassou, C., Basile, F., Poix, P., Michel, A.: Ann. Chim. *8,* 15 (1973)
157. Basile, F., Djega-Mariadassou, C., Poix, P.: J. Phys. Chem. Solids *35,* 1067 (1974)
158. Poix, P.: Thesis Paris (1963)
159. Nogues, M., Poix, P.: Ann. Chim. *3,* 335 (1968)
160. Banerjee, S. K., O'Reilly, W., Gibb, T. C., Greenwood, N. N.: Phys. Letters *20,* 455 (1966)
161. Banerjee, S. K., O'Reilly, W., Gibb, T. C., Greenwood, N. N.: J. Phys. Chem. Solids *28,* 1323 (1967)
162. Akimoto, S., Katsura, T., Yoshida, M.: J. Geomagnet. Geoelec. *9,* 165 (1957)
163. Chevallier, R., Bolfa, J., Mathieu, S.: Bull. Soc. Fr. Mineral. Cristallogr. *78,* 307 (1955)
164. Vanleerberghe, R., de Grave, E., Vandenberghe, R. E., Robbrecht, G. G.: J. Phys. (Paris) Colloque *41,* C1–179 (1980)
165. Melzer, K., Dehe, G., Mehner, H.: ibid *C1*-181 (1980)
166. Abe, M., Kawachi, M., Nomura, S.: J. Phys. Soc. Japan *31,* 940 (1971)
167. Monnier, G.: Bull. Soc. Chim. France 1252 (1959)
168. Rahmel, A., Jaeger, W., Becker, K.: Arch. Eisenhüttenw. *30,* 351 & 435 (1951)
169. Kazmanov, Y., Sorokin, L.: Tr. Uralsk Politekn. Inst. 154 (1961)
170. Ghose, J., Greenwood, N. N., Hallam, G. C., Read, D. A.: J. Solid State Chem. *11,* 239 (1974)
171. Abe, M., Kawachi, M., Nomura, S.: J. Phys. Soc. Japan *34,* 565 (1973)
172. Nakagawa, T.: ibid. *24,* 806 (1968)
173. Ramdani, A., Gleitzer, C., Gavoille, G., Cheetham, A. K., Goodenough, J. B.: J. Solid State Chem. (submitted)
174. Gerardin, R., Ramdani, A., Gleitzer, C., Gillot, B., Durand, B.: J. Solid State Chem. (in press)
175. Tellier, J.: Rev. Chim. Miné. *4,* 325 (1967)
176. Knoch, H., Dannheim, H.: Phys. Stat. Sol. *37a,* K135 (1976)
177. Kimura, T., Yoneda, M., Yamaguchi, T.: J. Am. Ceram. Soc. *60,* 180 (1977)
178. Rezlescu, N.: C.R. Acad. Sci. *268,* 136 (1969)
179. Dormann, J. L., Merceron, T., Renaudin, P., Brabers, V. A. M.: J. Phys. (Paris) Colloque *41,* C1–177 (1980)
180. Blasse, G.: Philips Res. Rep. Suppl. *3,* 1 (1964)
181. Rieck, G. D., Thijssen, J. J. M.: Acta Cryst. *B24,* 982 (1968)
182. Sawatsky, G. A., van der Woude, F., Morrish, A. H.: J. Appl. Phys. *39,* 1204 (1968)
183. Satya Murthy, N. S., Natera, M. G., Begum, R. J., Youssef, S. I.: Proc. Int. Conf. Ferrites, Kyoto, 60 (1970)
184. Jonker, G. H.: J. Phys. Chem. Solids *9,* 165 (1965)
185. Franke, H., Rosenberg, M.: J. Magn. Magn. Mater. *4,* 186 (1977)
186. Murray, P. J., Linnett, J. W.: J. Phys. Chem. Solids *37,* 619 (1976)
187. Yamzin, I. I., Belov, N. V., Nozik, Y. Z.: J. Phys. Soc. Japan *17,* Suppl. BIII, 55 (1962)
188. Palmer, W.: J. Appl. Phys. *33,* 1201 (1962)
189. Simsa, Z.: Czech. J. Phys. *B15,* 435 (1965)
190. Lotgering, F. K.: J. Phys. Chem. Solids *25,* 95 (1964)
191. Lotgering, F. K., van Diepen, A.: ibid. *34,* 1369 (1973)
192. Thackeray, M. M., David, W. I. F., Bruce, P. G., Goodenough, J. B.: Mater. Res. Bull. *18,* 461 (1983); *17,* 785 (1982)

193. Goodenough, J. B., Loeb, A. L.: Phys. Rev. *98,* 391 (1955)
194. Goodenough, J. B.: J. Phys. Soc. Japan *17,* Suppl. B-I, 185 (1962)
195. Ohnishi, I. H., Teranishi, T.: J. Phys. Soc. Japan *16,* 35 (1961)
196. Mexmain, J.: Ann. Chim. *6,* 297 (1971)
197. Kordes, E., Rottig, E.: Z. Anorg. Allg. Chem. *264,* 34 (1951)
198. Bertaut, E. F., Delorme, C.: C.R. Acad. Sci. *236,* 74 (1953)
199. Gorter, E. W.: Adv. Phys. *6,* 386 (1957)
200. Thery, J., Collongues, R.: C.R. Acad. Sci. *254,* 685 (1962)
201. Jefferson, C. F.: J. Appl. Phys. *36,* 1165 (1965)
202. Ohbayashi, K., Iida, S.: J. Phys. Soc. Japan *23,* 776 (1967)
203. Lenglet, M.: Rev. Chim. Miner. *5,* 1169 (1968)
204. Bergstein, A.: Mat. Res. Bull. *3,* 787 (1968)
205. Cervinka, L., Simsa, Z.: Czech. J. Phys. *B20,* 470 (1970)
206. Nagarajan, A., Agajanian, A. H.: J. Appl. Phys. *41,* 1642 (1970)
207. Dürr, J.: Thesis Lille 1980
208. Lerebourg, B., Dürr, J., D'Huysser, A., Bonnelle, J., Lenglet, M.: Phys. Stat. Sol. *61a,* K175 (1980)
209. Miyahara, S., Kino, Y.: Jap. J. Appl. Phys. *4,* 310 (1965)
210. Kitzinger, E., Simsa, Z.: Czech. J. Phys. *B18,* 955 (1968)
211. D'Huysser, A.: Thesis Lille 1982
212. Hannoyer, B.: Thesis Rouen 1983
213. Brabers, V. A. M.: Mat. Res. Bull. *18,* 861 (1983)
214. Lotgering, F. K., van Diepen, A. M.: J. Phys. Chem. Solids *38,* 565 (1977)
215. Strobel, P., Koltyberg, F. P., Wold, A.: J. Solid State Chem. *31,* 209 (1980)
216. Durif-Varanbon, A., Bertaut, E. F., Pauthenet, R.: Ann. Chim. (Paris) *13,* 526 (1956)
217. Blasse, G., Fast, J. F.: Philips Res. Rep. *18,* 393 (1963)
218. Miyahara, Y., Sai, F.: J. Phys. Soc. Japan *41,* 1522 (1976)
219. Robbins, M.: J. Phys. Chem. Solids *26,* 831 (1965)
220. Grandjean, F., Gerard, A.: Solid State Comm. *25,* 679 (1978)
221. Gleitzer, C., Cheetham, A. K.: 2nd Int. Conf. Solid State Chem. Eindhoven (1982)
222. Tricker, M. J., Vaishnara, P. P., Marshall, J. F., Donalson, J. D., Grimsey, R. H. A.: J. Inorg. Nucl. Chem. *43,* 95 (1981)
223. Van Diepen, A. M., Lotgering, F. K.: Solid State Comm. *27,* 255 (1978)
224. Lotgering, F. K.: J. Phys. Chem. Solids *35,* 1633 (1974)
225. Drofenik, M., Hanzel, D., Moljk, A.: J. Mater. Sci. *8,* 924 (1973)
226. Van Diepen, A. M., Lotgering, F. K.: J. Phys. Chem. Solids *35,* 1641 (1974)
227. Sauer, C., Köbler, U., Zinn, W., Stäblein, H.: ibid. *39,* 1197 (1978)
228. Aharoni, A., Schieber, M.: Phys. Rev. *123,* 807 (1961)
229. Hever, K. O.: J. Electrochem. Soc. *115,* 830 (1968)
230. Dudley, G. D., Steele, B. C. H., Howe, A. T.: J. Solid State Chem. *18,* 141 (1976)
231. Howe, A. T., Dudley, G. D.: ibid. 149
232. Dudley, G. D., Steele, B. C. H.: ibid. *21,* 1 (1977)
233. Howe, A. T., Dudley, G. D.: ibid. *30,* 157 (1979)
234. Katsura, T., Iwasaki, B., Kimura, S.: J. Chem. Phys. *47,* 4559 (1967)
235. Jette, E. R., Foote, F.: ibid. *1,* 29 (1933)
236. Roth, W. L.: Acta Cryst. *13,* 140 (1960)
237. Koch, F., Cohen, J. B.: ibid. *B25,* 275 (1969)
238. Cheetham, A. K., Fender, B. E. F., Taylor, R. I.: J. Phys. *C4,* 2160 (1971)
239. Catlow, C. R. A., Fender, B. E. F.: ibid. *C8,* 3267 (1975)
240. Gavarri, J., Berthet, C., Carel, C., Weigel, D.: C.R. Acad. Sci. *285C,* 237 (1977)
Gavarri, J., Berthet, C., Weigel, D.: ibid. *284C,* 325 (1977)
241. Koch, F., Fine, M. E.: J. Appl. Phys. *38,* 1470 (1966)
242. Battle, P. D., Cheetham, A. K.: J. Phys. *C12,* 337 (1979)
243. Wilkinson, C., Cheetham, A. K., Long, G. J., Battle, P. D., Hope, D. A. O.: Inorg. Chem. *23,* 3136 (1984)
244. Baüer, E.: Thesis Nancy (1979)
245. Elias, D. J., Linnett, J. W.: Trans. Faraday Soc. *65,* 2673 (1969)

246. Checherskaya, L. F., Romanov, V. P., Tatsienko, P. A.: Phys. Stat. Sol. *19a,* K177 (1973)
247. Greenwood, N. N., Howe, A. T.: J. Chem. Soc. Dalton Trans. 122 (1972)
248. Le Corre, G.: Thesis Nancy (1973)
249. Gleitzer, C.: Mat. Res. Bull. *15,* 955 (1980)
250. Gleitzer, C.: ibid. *15,* 507 (1980)
251. Gourdin, W. H., Kingery, W. D.: J. Mater. Sci. *14,* 2053 (1979)
252. El Balkhi, M., Zanne, M., Gleitzer, C., Courtois, A.: J. Solid State Chem. *18,* 293 (1976)
253. El Balkhi, M., Courtois, A., Zanne, M., Gleitzer, C.: C.R. Acad. Sci. *285C,* 129 (1977)
254. El Balkhi, M.: Thesis Nancy (1977)
255. Cirilli, V., Burdese, A.: Metall. Ital. *44,* 371 (1952)
256. Schenck, R., Franz, H., Willecke, H.: Z. Anorg. Allg. Chem. *184,* 16 (1929)
257. Holmquist, S. B.: Nature *185,* 604 (1960)
258. Borgiani, C.: 7èmes Journées Int. Sidérurgie, Versailles, oct. (1978)
259. Malaman, B., Alebouyeh, H., Jeannot, F., Courtois, A., Gerardin, R., Evrard, O.: Mat. Res. Bull. *16,* 1139 (1981)
260. Malaman, B., Alebouyeh, H., Courtois, A., Gerardin, R., Evrard, O.: Mat. Res. Bull. *17,* 795 (1982)
261. De Sitter, J., Govaert, A., de Grave, E., Chambaere, D., Robrecht, G. G.: Phys. Stat. Sol. *43a,* 619 (1977)
262. Buffat, B., Demazeau, G., Pouchard, M., Fournes, L., Dance, J. M., Fabritchnyi, P., Hagenmuller, P.: C.R. Acad. Sci. *292,* Ser. II, 509 (1981)
263. Wollan, E. O., Koehler, W. C.: Phys. Rev. *100,* 545 (1955)
264. Goodenough, J. B., Longo, J. M.: Landolt-Börnstein New Series, Vol. *III/4a,* 126, Springer Verl., Berlin 1970
265. Callaghan, A., Moeller, C. W., Ward, R.: Inorg. Chem. *5,* 1572 (1966)
266. Takeda, Y., Naka, S., Takano, M., Shinjo, T., Takada, T., Shimada, M.: Mat. Res. Bull. *13,* 61 (1978)
267. Takeda, T., Yamaguchi, Y., Watanabe, H.: J. Phys. Soc. Japan *33,* 967 (1972)
268. Oda, H., Yamaguchi, Y., Takei, H., Watanabe, H.: J. Phys. (Paris) Colloque *38,* C1–121 (1977)
269. Takeda, T., Komura, S., Fugii, H.: J. Magn. Magn. Mater. *31–34,* 797 (1983)
270. McChesney, J. B., Sherwood, R. C., Potter, J. F.: J. Chem. Phys. *43,* 1907 (1965)
271. Takano, M., Nakanishi, N., Takeda, Y., Naka, S., Takada, T.: Mat. Res. Bull. *12,* 923 (1977)
272. Raccah, P. M., Goodenough, J. B.: Phys. Rev. *155,* 932 (1967)
273. Bartlett, N., Rao, P. R.: Proc. Chem. Soc. 393 (1964)
274. Takeda, Y., Naka, S., Takano, M.: J. Phys. (Paris) Colloque *40,* C2–331 (1979)
275. Takano, M., Kawachi, J., Nakanichi, N., Takeda, Y.: J. Solid State Chem. *39,* 75 (1981)
276. Komornicki, S., Fournes, L., Grenier, J. C., Menil, F., Pouchard, M., Hagenmuller, P.: Mat. Res. Bull. *16,* 967 (1981)
277. Goodenough, J. B., Kafalas, J. A., Longo, J. M.: in Preparative Methods in Solid State Chemistry (Hagenmuller, P., ed.), Academic Press, N.Y. 1972, p. 1
278. Nomura, S.: Landolt-Börnstein New Series, Vol. *III/12a,* 368, Springer Verl., Berlin 1978
279. Lucchini, E., Meriani, S., Minichelli, D.: Acta Cryst. *B29,* 1217 (1973)
280. Malinofsky, W. W., Kedesdy, H.: J. Am. Chem. Soc. *76,* 3090 (1954)
281. Erchak, M., Fankuchen, I., Ward, R.: ibid. *68,* 2085 (1946)
282. Scholder, R., Bunsen, H. V., Zeiss, W.: Z. Anorg. Allg. Chem. *283,* 330 (1956)
283. Scholder, R., Kindervater, F., Zeiss, W.: ibid. 338
284. Van Hook, H. J.: J. Phys. Chem. *68,* 3786 (1964)
285. Zanne, M., Gleitzer, C., Aubry, J.: Bull. Soc. Chim. France, 2451 (1971)
286. Ichida, T.: J. Solid State Chem. *7,* 308 (1973)
287. Zanne, M., Gleitzer, C.: Bull. Soc. Chim. France, 1567 (1971)
288. Takeda, Y., Shimada, M., Kanamaru, F., Koizumi, M., Yamamoto, N.: Mat. Res. Bull. *9,* 537 (1974)
289. Takeda, Y., Shimada, M., Kanamaru, F., Koizumi, M.: J. Solid State Chem. *7,* 229 (1973)
290. Shimada, M., Takeda, Y., Taguchi, H., Kanamaru, F., Koizumi, M.: J. Cryst. Growth *29,* 75 (1975)
291. Gleitzer, C., Zanne, M.: C.R. Acad. Sci. *267C,* 1601 (1968)

292. Aubry, J., Gleitzer, C., Zanne, M.: ibid. *268C,* 2173 (1969)
293. Zanne, M.: Thesis Nancy (1972)
294. Mori, S.: J. Phys. Soc. Japan *28,* 44 (1970)
295. Gleitzer, C., Zanne, M., Zeller, C.: C.R. Acad. Sci. *270B,* 1496 (1970)
296. MacChesney, J. B., Potter, J. F., Sherwood, R. C., Williams, H.: J. Chem. Phys. *43,* 3317 (1965)
297. Zanne, M., Gleitzer, C.: J. Solid State Chem. *6,* 163 (1973)
298. Evrard, O., Malaman, B., Jeannot, F., Tannieres, N., Aubry, J.: C.R. Acad. Sci. *278C,* 241 (1974)
Malaman, B., Evrard, O., Tannieres, N., Aubry, J., Courtois, A., Protas, J.: Acta Cryst. *B31,* 1310 (1975)
299. Kato, K., Kawada, I., Kimizuka, N., Katsura, T.: Z. Krist. *141,* 314 (1975)
300. Kimizuka, N., Kato, K., Shindo, I., Kawada, I., Katsura, T.: Acta Cryst. *B32,* 1620 (1976)
301. Sugihara, T., Siratori, K., Shindo, I., Katsura, T.: J. Phys. Soc. Japan *45,* 1191 (1978)
302. Sekine, T., Katsura, T.: J. Solid State Chem. *17,* 49 (1976)
303. Kimizuka, N., Katsura, T.: ibid. *15,* 151 (1975)
304. Kimizuka, N., Katsura, T.: ibid. *13,* 176 (1975)
305. Katsura, T., Sekine, T., Kitayama, K., Sugihara, T., Kimizuka, N.: ibid. *23,* 43 (1978)
306. Gerardin, R., Alebouyeh, A., Jeannot, F., Courtois, A., Malaman, B., Evrard, O.: Mat. Res. Bull. *15,* 647 (1980)
307. Tannieres, N., Evrard, O., Jeannot, F., Aubry, J.: Rev. Chimie Miné. *12,* 518 (1975)
308. Tannieres, N.: Thesis Nancy (1978)
309. Gerardin, R., Alebouyeh, A., Malaman, B., Evrard, O.: Rev. Chimie Miné. *18,* 625 (1981)
310. Nakagawa, Y., Inazumi, M., Kimizuka, N., Siratori, K.: J. Phys. Soc. Japan *47,* 1369 (1979)
311. Inazumi, N., Nakagawa, Y., Tanaka, N., Kimizuka, N., Siratori, K.: ibid. *50,* 438 (1981)
312. Aqachmar, H.: Thesis Nancy 1982
313. Akimitsu, J., Inada, Y., Siratori, K., Shindo, I., Kimizuka, N.: Solid State Comm. *32,* 1065 (1979)
314. Nayaransamy, A., Muthukumarasamy, P., Kimizuka, N., Nagarajan, T.: Proc. Nucl. Phys. Sol. Phys. Symp. *21C,* 574 (1978)
315. Tanaka, N., Kato, M., Kimizuka, N., Siratori, K.: J. Phys. Soc. Japan *47,* 1737 (1979)
316. Sugihara, T., Abe, M., Katsura, T.: ibid. *40,* 1211 (1976)
317. Modaressi, A., Courtois, A., Gerardin, R., Malaman, B., Gleitzer, C.: J. Solid State Chem. *40,* 301 (1981)
318. Iraldi, R., le Caër, G., Gleitzer, C.: Solid State Comm. *40,* 145 (1981)
319. Venturini, G., Courtois, C., Steinmetz, J., Gerardin, R., Gleitzer, C.: J. Solid State Chem. *53,* 1 (1984)
320. Smuts, J., Steyn, J. G. D., Boyeyens, J. C. A.: Acta Cryst. *25B,* 1251 (1969)
321. Gleitzer, C.: Int. Conf. Basic Prop. Binary Oxides, Sevilla 1983 (Dominguez-Rodriguez et al., Ed.) Univ. Sevilla 1984
322. Coey, J. M. D., Monkarika, D., McDonagh, C. M.: Solid State Comm. *41,* 797 (1982)
323. Slonczewski, J.: in Magnetism: A Treatise on Modern Theory and Materials (Rado, G. T., Suhl, H., eds.), Acad. Press N.Y. 1963, Vol. 1, chap. V
324. Gyorgy, E. M., Dillon, J. F., Remeika, J. R., Remeika, J. P.: IBM J. Res. & Dev. *14,* 321 (1970)

An Alternative Approach to Non-Molecular Crystal Structures
With Emphasis on the Arrangements of Cations

M. O'Keeffe[1] **and B. G. Hyde**[2]

1 Department of Chemistry, Arizona State University, Tempe, Arizona, U.S.A.
2 Research School of Chemistry, Australian National University, Canberra, A.C.T., Australia

The important thing in science is not so much to obtain new facts as to discover new ways of thinking about them. *W. L. Bragg*[105)]

The alternative approach to structures is an inversion of the traditional one: we concentrate on a packed array of cations, into the interstices of which the anions are inserted, rather than putting cations into a packed anion array. Even in conventional terms the logic of the new approach can be justified, since many structures have their own antitypes (and even more show a partial structure/antistructure relation).

In this way we find that, in oxides for example, many cation arrangements are identical to the arrangements of the atoms in known (or plausible) alloy structures. (In some cases the cations and atoms are identical in the two cases.) This is particularly helpful in those oxide examples where it has not previously been possible to describe the structure in *any* simple terms, often because the anion array is not regular in any simple way, for instance in many metal sulphates. But it is also revealing in some cases that are describable in conventional terms, e.g. the humites. A long (but still incomplete) list of examples is given in Table 3. Sufficient specific cases are described and discussed in detail in the text in order to expose the principles and some of the advantages of this unfamiliar approach.

One of these advantages is that, in favourable cases, it leads at once to complete quantitative descriptions of the structures (i.e. until cell dimensions and anion position parameters), e.g. for spinel, garnet, bixbyite etc. Another is that it throws a new light on possible martensitic mechanisms for phase transformations (e.g. β- → γ-Ca_2SiO_4, low- → high-temperature K_2SO_4), and on the possibility of anomalous substitution (e.g. of CO_3^{2-} for PO_4^{3-} in apatite). It also reveals an explanation for those phase transitions, due to high pressure, that are *not* attended by an increase in primary coordination number, e.g. olivine → spinel, zircon → scheelite.

1 Introduction

For more than sixty years, the basic approach to the crystal chemistry of "ionic" structures, particularly those of the solid metal oxides, has placed almost exclusive emphasis on the geometry of the anion array. Consideration of the cations is normally limited to a discussion of their immediate environment, i.e. to bond lengths and coordination polyhedra.

Although it has often proved useful as a mnemonic, this approach has led to a number of misconceptions about relative atomic sizes and the origin of close-packing geometry, to some of which we allude below. More relevant in the present context is the observation that one natural and simple description of crystal structures has been overlooked, and an unnecessarily complicated and opaque one used instead. We will provide many examples throughout this article.

The emphasis here is on providing a useful and accurate description of crystal structures. It is one complementary to the traditional one in that the emphasis will be on the cation packing and their coordination around the anion (usually oxygen but sometimes also nitrogen and fluorine)[1]. The emphasis is on the purely geometrical aspects of crystal structures and is thus independent of any theory of bonding in solids. Indeed occasional use of the terms *cation* and *anion* in this paper is just for convenience in distinguishing between more electropositive and more electronegative atoms and does not represent a commitment to an ionic description of bonding.

On the other hand, we would be remiss not to state briefly our reason for thinking that our new approach might be more appropriate physically. The main reason is simply that in oxides such as silicates the "cations" such as silicon are larger *(as measured by the extent of short range, Pauli, forces)* than the "anions". We[1,2)] have recently discussed the evidence for and some of the implications of this observation. In particular we find it essential to understanding metal-oxygen-metal bond angles, coordination numbers and volume changes in phase transitions.

We feel that it is very important *not* to draw conclusions about bonding just from the observation of the occurrence of a particular structure. It is well known that the same structure can serve for crystals of presumably different bond types: Mg_2Sn and Li_2O, CuZn and CsCl are examples of pairs of compounds with the same well-known structures. Less well-known perhaps but more important in the present context is the wide occurrence of anti-structures in which cation and anion positions are interchanged. In Table 1 a number of structure-antistructure pairs is given, mainly to emphasize how wide-spread the phenomenon is, and that it encompasses both simple and complex structures.

We do not give explicit references to structure types here, or often elsewhere. Most of the structures discussed can be found in the standard reference texts[3–5)] or in *Structure Reports*. We have considered a few of the structure types appearing in Table 1 in some detail elsewhere[6,7)].

Also of interest is the occurrence of the same partial structure as both an anion and a metal atom array even when two complete structures are not antistructures of each other.

1 It is not necessary to exclude sulphur and heavier anions. But, at this juncture, it is simpler to do so

Table 1. Examples of structure and antistructure

Structure[a]	Compound with structure	Compound with antistructure
B 11	PbO(E)	OLi(H)
C 1	ThO_2	OLi_2
C 1	PbF_2	$SiMg_2$
C 4	TiO_2	NTi_2, δ-NCo_2
$E0_1$	PbFCl	SiMgBa
C 23 = C 37	$PbCl_2$	$SiBa_2$
C 6	CdI_2	FAg_2
C 19	$CdCl_2$	NCa_2, OCs_2
–	α-PbO_2	ζ-NFe_2
–	γ-MnO_2	ζ-CNb_2
–	$NaFeO_2$	$PICa_2$
$D0_3 = D0_{11}$	YF_3	CFe_3
$D0_9$	ScF_3	NCu_3
$D0_6 = D0_{18}$	LaF_3	$AsNa_3$
–	$PuBr_3$	BRe_3
$D5_2$	La_2O_3	Sb_2Mg_3
$D5_3$	Y_2O_3	N_2Mg_3
$D5_{10}$	α-La_2S_3	C_2Cr_3
–	α-U_2S_3	P_2Hf_3
$D7_3$	Th_3P_4	As_3Ti_4
–	U_3S_5	Ge_3Rh_5
–	La_2O_2S	Si_2Al_2Ca
C 22	K_2ThF_6	$Si_2BNi_6(PFe_2)$
–	Tl_3PbCl_5	Si_4Zr_5
$E2_1$	$CeCrO_3$	$GaNCr_3$
–	$CaIrO_3$	$GeCCr_3$
–	K_2NiF_4	Sb_2OCa_4

[a] Strukturbericht designations

Table 2. Examples of partial antistructure in which anions and metal atoms have the same arrangement[a]

Structure	Antistructure
$CaTiO_3 = (Ti)CaO_3$	CFe_3
$KFeO_2 = (Fe)KO_2$	$MgCu_2$
$Ca_3Al_2Si_3O_{12} = (Si_3Al_2)Ca_3O_{12}$	$Rh_3Bi_{12} = RhBi_4$
$Bi_4Si_3O_{12} = (Si_3)Bi_4O_{12}$	$As_4Cu_{12} = Cu_3As$
$RbAg_4I_5 = (Ag_4)RbI_2I_3$	CAl_2Mo_3
$CaCu_3Mn_4O_{12} = (Cu_3Mn_4)CaO_{12}$	WAl_{12}
$U_2Fe_2S_5 = (Fe_2)U_2S_5$	C_2Fe_5
$Ba_3Fe_3S_7 = (Fe_3)Ba_3Se_7$	B_3Ru_7
Sb_2O_3	$Fe_3W_3C = (FeC)Fe_2W_3$
$La_2Ti_2O_7 = (O)La_2Ti_2O_6$	$Fe_3W_3C = (Fe_4)Fe_2C_2W_6$
HTB e.g. $(Rb)W_3O_9$	$W_9Co_3C_4 = (Co_3C)C_3W_9$

[a] The atoms not in parentheses in the right hand side of both columns have the same structure

Table 2 records some examples of this phenomenon in which *oxygen* arrays in oxides are the same as metal atom arrays in alloys. Recognition of this fact has been exploited to simplify the description of complex alloys (see especially Andersson[8)]), which is essentially the reverse of what we propose to do here, namely to simplify the description of oxide structures by giving them in terms of known alloy structures. Nevertheless Tables 1 and 2 provide striking evidence of Nature's parsimony in the use of patterns in crystal structures.

The sphere close-packing geometry (we prefer the term *eutaxy*[9)] in this context) is of course of very frequent occurrence for both cation and anion arrays in binary compounds, and the description of structures in terms of the filling of the interstices of these arrays is both familiar and useful. However, when ternary etc. compounds are considered, one has to deal mainly with examples with mixed cations and *one* type of anion (e.g. $MgAl_2O_4$) rather than the reverse (e.g. Si_2N_2O). This means that the cation array is now a packing of two (or more) kinds of atom. Remarkably, in too many instances for it to be entirely a coincidence, it is found that the *cation arrays in such oxides often correspond to (binary) alloy structures.* There are some familiar examples: the identity of the $MgAl_2$ array in $MgAl_2O_4$ (spinel) to the structure of the cubic Friauf-Laves phase $\mathbf{MgCu_2}$[2] is well known. We shall adduce many other examples in what follows.

In a number of symmetrical structures with few free parameters (e.g. those of spinel, garnet and pyrochlore), it is shown that bond-length considerations alone are sufficient to determine these parameters. In a further group of structures, fixing the bond lengths is not sufficient to determine all the parameters. One must allow the structures to relax to minimise the non-bonded repulsions. For structures with just a few remaining degrees of freedom (e.g. rutile, bixbyite and calcite), minimising the non-bonded repulsions can be simulated by maximising the volume subject to the constraint of fixed bond length[9)]. (In the first group of structures the high-symmetry configuration is already a maximum-volume one.) In all the examples cited, the agreement between observed and calculated structural parameters is good.

2 Oxides as Stuffed Alloys

All the cation arrays we will discuss in the rest of this paper are either the structures of alloys (including of course the elements) or very simply related to them. They are summarized in Table 3. Some, such as MgO and ThO_2, are sufficiently well-known to need little comment except perhaps to remark that they represent the structures of a large number of individual compounds. Others (those marked with an asterisk in Table 3, to many of which the same remark applies) are singled out for further discussion because the new description sheds new light on incompletely understood structures or, as in the case of the sulphates, because our approach provides the first satisfactory description of the structure.

There are of course very many structures that appear not to be based on known alloy structure types.

---

2 Chemical formulae in heavy type indicate structure types, not compounds

Table 3. Examples of oxide, nitride and fluoride structures in which the cation arrays are alloy structures

I. Cation arrays derived from cubic eutaxy	
Cu (A 1)	MgO, ThO_2, Y_2O_3, PbO, Cu_2O[a], Cu_3O_4[a], CuO[a]
Cu_3Au (L 1_2)	K_3SiF_7, Ba_3SiO_5, $TlPd_3O_4$, MPt_3O_6 with M = Mn, Co, Zn, Mg, Ni
CuPt (L 1_1)	α-$NaFeO_2$
$MoPt_2$	Li_2CuO_2
WAl_5	Li_5ReO_6
$TiAl_3$ (D 0_{22})	Na_3PaF_8
AB	$CaWO_4$[a]
AB	$La_2Ti_2O_7$, $AgSbO_3$[a]
II. Other eutactic cation arrays	
Mg (A 3)	La_2O_3[a], ZnO, YF_3, LaF_3
Sm = $[(hc)h]$	Th_3N_4[a], $YbFe_2O_4$
$[(hc)^2h]$	$Yb_2Fe_3O_7$
$[(hc)^3h]$	$Yb_3Fe_4O_{10}$
$[(hc)^4h]$	$Yb_4Fe_5O_{13}$
$BaPb_3$	Ca_3SiO_5
Ni_3Sn (D 0_{19})	Li_3PO_4, K_3NbF_7OH
III. Cation arrays derived from body-centred cubic packing	
W (A 2)	χ-La_2O_3 (high T)
β-Hg	TiO_2 (rutile)[a], $ZrSiO_4$[a]
NaTl (B 32)	$KFeO_2$, KH_2PO_4
CuZn (B 2)	$SrTiO_3$[a], $Ca_2Fe_2O_5$, $TlSbF_6$, $KAlF_4$
$MoSi_2$ (C 11_b)	Sr_2TiO_4, Bi_2SiO_4, Sb_2ZnO_6, La_2MoO_6, Bi_2MoO_6, $CsSb_2F_7$
Fe_3Al (L 2_1)	BaW_3O_6, Na_3PO_4
UCo	$K_2Pb_2O_3$
Cu_2MnAl (L 2_1)	K_2NaAlF_6
IV. Cation arrays based on other alloy structures	
$MgCu_2$ (C 15)	Mg_2AlO_4[a]
CrB (B 33)	$CsBeF_3$[a], $KClO_3$[a], $BaGeO_3$, β-Ga_2O_3, α-NaOH
FeB (B 27)	$BaSO_4$[a]
Mn_5Si_3 (D 8_8)	$Ca_5P_3O_{12}OH$ (apatite)[a], $Y_5Si_3O_{12}N$, $Ba_5Re_3O_{15}X$[a] with X = F, $(CO_3)_{1/2}$ or O_2
NiAs (B 8_a)	$CrVO_4$, $CuSO_4$, $BaNiO_3$, $CaCO_3$ (aragonite)
A_2B_5[b]	$Na_5Zr_2F_{13}$
AB_3[b]	Na_3TaF_8
Th_3P_4 (D 7_3)	$Bi_4Si_3O_{12}$, $K_2Mg_2S_3O_{12}$
W_2CoB_2	kotoite $Mg_3(BO_3)_2$[a], $Cd_3GeO_4(OH)_2$
U_3Si_2 (Mo_2CoB_2)	$Ca_2ZnSi_2O_7$, melilite $(Ca, Na)_2(Mg, Al, Si)_3O_7$, $Ba_2TiSi_2O_8$, $K_2V_3O_8$

Table 3 (continued)

Ni_2Si (C 37 ≈ C 23)	β-Ca_2SiO_4[a], β-K_2SO_4[a], Ba_2TiO_4, $FeCoBO_4$, $CaFe_2O_4$, K_2NbF_7, α-$Zr(SO_4)_2$
Ni_2In (B 8_b)	Mg_2SiO_4 (olivine)[a], K_2CO_3, $LiMgPO_4$, γ-Ca_2SiO_4[a]
FeS_2 ($PtBi_2$) (C 2_a)	ZrP_2O_7, SiP_2O_7
$CuAl_2$ (C 16)	$Pb_3O_4(Pb_2^{2+}Pb^{4+}O_4)$, $ZnSb_2O_4$, $Ca_2F_2CO_3$ (brenkite)
ω-phase	$BaNb_2O_6$, $PbSb_2O_6$, K_2PtF_6, Sr_2PbO_4
Cr_3Si (A 15)	Ag_3PO_4[a], $NaPt_3O_4$[a], $Ca_3Al_2Si_3O_{12}$ (garnet)[a]
Re_3B	$Mg_2SiO_4 \cdot Mg(OH,F)_2$ (norbergite)[a], Fe_3BO_6
$PtPb_4$	$BaMn_4O_8$ (hollandite)
Th_3Pd_5	$K_3V_5O_{14}$
~ Mo_3CoSi	vesuvianite[a], $[Ba_{0.64}(H_2O)_{1.36}]Mn_5O_{10}$ psilomelane
V. Some oxides with tetrahedral frameworks[a]	
Si (A 4)	SiO_2 (cristobalite)
Lonsdaleite	SiO_2 (tridymite)
γ-Ge	SiO_2 (keatite)
AlP	$AlPO_4$
LiAlSi	$NaAlSiO_4$ (carnegieite)
LiGaGe	$KAlSiO_4$ (kalsilite)
CrB_4	$CaAl_2Si_2O_8$ (monoclinic form)
$CeCu_2$	$RbAlSiO_4$
K_4Si_{23}	$(N_2)_3CH_4(SiO_2)_{23}$ (melanophlogite)

[a] Compounds discussed in some detail in the text
[b] hypothetical alloy type

2.1 Some Structures Derived by Filling the Interstitial Sites in Eutactic Arrays of Cations

2.1.1 Tetrahedral Sites of Cubic Eutaxy

2.1.1.1 The Copper Oxides

There are three well-established oxides of copper containing Cu^I and/or Cu^{II} *viz.*: Cu_2O, Cu_4O_3 and CuO. Their structures are all based on a cubic-close packing (c.c.p.) of copper (somewhat distorted in Cu_4O_3 and CuO) with oxygen inserted into tetrahedral sites in such a way as to be a solution to the problem of having Cu^I in linear two-fold coordination and Cu^{II} in planar four-fold coordination. These have been discussed from this point of view elsewhere[10)], and it has been speculated that the observed structures are, in fact, the only solutions to the problem.

2.1.1.2 The Bixbyite ($D5_3$) Structure

The bixbyite structure of e.g. Sc_2O_3 represents a solution to the problem of filling three quarters of the tetrahedral sites of cubic close-packed Sc with O. It is not the simplest solution; that would be the structure obtained by filling the tetrahedral sites not filled in Cu_2O (i.e. the pyrochlore structure of $AgSbO_3$, with Ag, Sb replaced by Sc, cf. Sect. 2.1.1.3). With normal Sc-O bond lengths that structure would have short O...O distances, and so an alternative structure that allows adjustment of the O...O and M...M distances is adopted.

In the bixbyite structure there are two crystallographically distinct Sc atoms. Taken together Sc(1) and Sc(2) are arranged as (Au and Cu) in Cu_3Au so that the tetrahedral sites are $Sc(1)Sc(2)_3$. Three quarters of these are filled in such a way that the empty sites fall on non-intersecting three-fold axes (rather than on intersecting axes, as in the simpler structure described above). In Sect. 3.5 we discuss the rather substantial (~0.3 Å) displacements of atoms from these ideal positions that occur in the real structure. It might be mentioned here that it is *not* correct to describe the displacements as oxygen moving off-centre in the Sc_4 tetrahedron. Rather, the tetrahedron distorts and oxygen remains having four very nearly equidistant neighbours.

2.1.1.3 The Pyrochlore Structure

Pyrochlore, $(Na, Ca)_2(Nb, Ti)_2(O, F)_7$ is one of a large group of cubic minerals that includes atopite, $(Ca, Mn, Na)_2Sb_2(O, OH, F)_7$; and indeed the group is sometimes referred to as atopites.

The simplest pyrochlores, which we consider here, can be written $A_2B_2O(1)_6O(2)$ and typical compounds might be $Y_2Ti_2O_7$ or $Ca_2Sb_2O_7$, although there are variants such as $AgSbO_3$ in which the O(2) site is not occupied. There are also closely related compounds such as $RbNbTeO_6$. Elsewhere[7] we have discussed in particular the arrangement of the O(1) sites, and its occurrence in both oxides and alloys. Here we focus attention on the cation arrays and the anion coordination.

The symmetry of the structure is $Fd3m$. A atoms are in 16(*c*): 0, 0, 0 etc.; B atoms are in 16(*d*): 1/2, 1/2, 1/2 etc.; O(1) in 48(*f*): *x*, 1/8, 1/8 etc.; and O(2) in 8(*a*): 1/8, 1/8, 1/8 etc. It will be noticed that the structure is analogous to that of spinel (Sect. 2.8) in that there are two parameters, *x* and the unit cell edge *a*, required to completely specify the structure. However, there is an interesting difference that we point out below.

Both the A atom array and the B atom array are identical to that of the Al atoms in $MgAl_2O_4$ (spinel) and to that of the Cu atoms in $MgCu_2$. Taken together they are in cubic eutaxy, and represent a superstructure of face-centered cubic (with consequently a doubled unit cell edge). The tetrahedral sites of this array are, per A_2B_2 formula unit, six A_2B_2, one A_4 and one B_4. The O(1) atoms are in the A_2B_2 tetrahedra, which they would centre if $x = 3/8$. [It is worth noting in passing that because the remaining tetrahedral sites, those *not* occupied by O(1), are those occupied in Cu_2O, the cuprite and pyrochlore structures such as $AgSbO_3$ are, in this sense, complementary.] The O(2) atoms centre the A_4 tetrahedra.

2.1.2 *Tetrahedral and/or Octahedral Sites of Cubic, Hexagonal and Mixed Eutaxies*

2.1.2.1 Fluorite (C 1) and NaCl (B 1)

For later use in this section it is sufficient to recall that both these structures are composed of cations in cubic eutaxy. In the fluorite type the anions occupy all the tetrahedral interstices in the cation array while, in the rock salt structure they occupy all the octahedral interstices. All the structures considered in this section are lamellar intergrowths of these two types (or of c.c.p. cations and B 1).

2.1.2.2 A-Type La_2O_3 (D 5_2)

In this structure the cation has a coordination number of seven. Best regarded as a monocapped octahedron, the coordination polyhedron may also be viewed as a cube of eight anions with a slight rhombohedral distortion. The cation is off-centre (in both descriptions) with the result that the eighth oxygen of the cube is too distant to be included in the coordination figure[11)]. But, in these terms, the structure is still rather puzzling; for example the La–O bonds fall into two groups, four short (~ 2.4 Å) and three long (~ 2.7 Å), a point to which we will return later.

Previous authors have pointed out[12, 13)] that the description of this structure in terms of *anion*-centred polyhedra is very straightforward: it is simply a hexagonal eutactic array of cations (slightly distorted, alternate layer spacings being 3.26 and 2.86 Å) with oxygens between the (0001) layers of cations occupying, alternately, all the tetrahedral or all the octahedral sites[3]. The first layer is, of course, a $\{111\}_{f.c.c.}$ layer of fluorite type, and the second a $\{111\}_{f.c.c.}$ layer of **NaCl** type. Hence, the A-type structure is an intergrowth of alternate, single layers of C 1 and B 1. This description is depicted in Fig. 1.

The stacking sequence of cation layers is clearly a consequence of anion . . . anion repulsion: any different sequence of the eutactic layers would result in face-sharing between (at least some) OLa_6 octahedra and OLa_4 tetrahedra[4]. As it is, only *edge*-sharing occurs: given the La_2O_3 stoichiometry and the interstices occupied, the anions are as far apart as possible. The anion layer sequence is then inevitably . . . *abcabc* . . . . (But note that the anion layer spacings are *not* those of cubic eutaxy.) This "avoidance principle" appears to determine the cation layer-stacking sequences in all the structures in this section.

The two groups of bond lengths are readily explained by an inversion of Pauling's rule: in the OLa_6 octahedron the O–La bond order is 2/6 = 1/3; in the OLa_4 tetrahedron it is 2/4 = 1/2. The former are therefore weaker and longer (2.72 Å) and the latter are

3 Hence the alternating layer spacings: the larger one corresponds to twice as many oxygens as the smaller. But cf. also the bond length discussion below

4 Between adjacent layers in hexagonal eutaxy both octahedra and tetrahedra share faces (with their own kind) whereas in cubic eutaxy tetrahedra share faces only with octahedra, and *vice versa*. The latter situation also obtains within layers (in both kinds of eutaxy): then octahedra share only edges with other octahedra, and similarly tetrahedra share only edges with each other

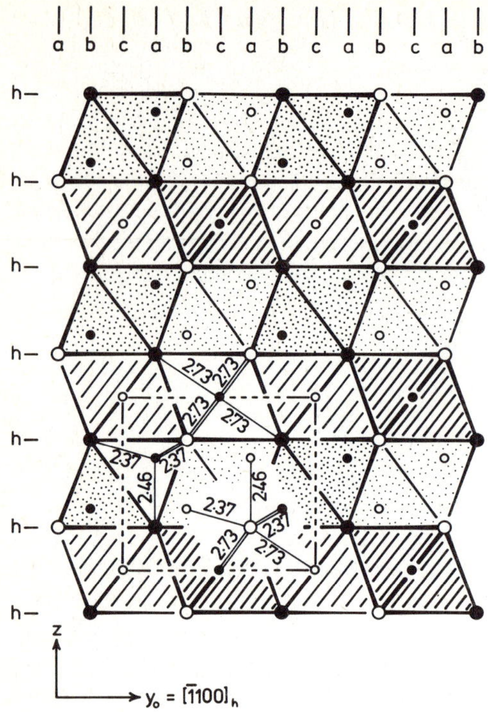

Fig. 1. The A-type La_2O_3 structure projected on (11$\bar{2}$0): all atoms at one of two heights, 0 or 1/2 (*open* and *filled circles*); *large circles* = La, *small circles* = O. The projected unit cell is outlined, and the seven shortest La–O bonds shown for one La atom. The polyhedra are anion-centred, OLa_6 octahedra and OLa_4 tetrahedra. Note the ... αβαβ ... (*h*) sequence of cation layers and the ... *abcabc* ... sequence of anion layers. The former are approximately eutactic, the latter are not

stronger and shorter (3 of 2.38 Å, and 1 of 2.45 Å to the "apical" La – a common deformation of hexagonal structures with tetrahedral coordination, e.g. wurtzite[14]).

Isostructural compounds include some other lanthanide and actinide sesquioxides and oxide-chalcogenides such as La_2O_2X (X = S or Se) as well as Th_2N_2X (X = O, S, Se), Th_2NOX (X = P, As) and analogous uranium compounds. In all cases the third anion is the one in octahedral coordination. (Antistructures include N_2Li_2Zr.)

{If the odd anion is too large for octahedral coordination, then two related structures are observed:

(i) the Sc_2O_2S type in which the S atom is in *trigonal prismatic* coordination; the Sc_2O_2 layers being as in La_2O_3, i.e. trigonal layers of C1 type. The cation layer sequence is now ...αββααββα..., and the Sc_2S layers are (0001) layers of **WC**- or **NiAs**-type.

(ii) The La_2O_2Te type in which Te atoms are in La_8 cubes, forming a unit {100} layer of CsCl(B2) type. Of necessity the La_2O_2 layers are now tetragonal, *viz.* $\{100\}_{f.c.c.}$ layers of C1 [as in (red) tetragonal PbO]. The $TeLa_2$ part is **MoSi₂** type. Compounds M_2N_2Te and M_2NOX (M = Th, U and X = Sb, Bi) are isostructural. Antistructures include P_2Zn_2Ba.}

2.1.2.3 Th_3N_4

Like **A-La₂O₃** this structure[15] also has a eutactic cation array, but now the layer sequence is *hhc* = ...αβα(βγβγαγαβα)βγβ..., a nine-layer sequence instead of the *h* = ...αβαβ... two layer sequence of A-La_2O_3. The reason is the same "avoidance principle".

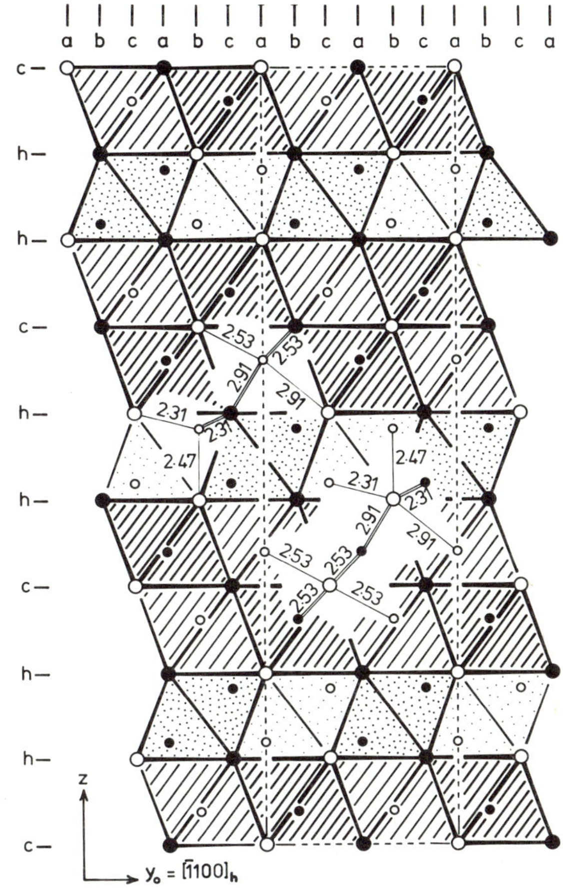

Fig. 2. The structure of Th_3N_4 projected on (11$\bar{2}$0): all atoms at one of two heights, 0 or 1/2 (*open* and *filled circles*): *large circles* = Th, small = N. The projected hexagonal unit cell is outlined, and the bonds to the two types of Th atoms (ThN_6 and ThN_7) are shown. Polyhedra are NTh_4 and NTh_6. Note cation and anion layer sequences

As La_2O_3 = LaO_2^-(C 1) + LaO^+(B 1) so Th_3N_4 = ThN_2^{2-}(C 1) + 2 ThN^+(B 1); i.e. *single* layers of C 1 alternate with *double* layers of B 1. The cation layer between the two different layer types must be *h*, to prevent NTh_6 octahedra sharing faces with NTh_4 tetrahedra, but the cation layer between the two B 1 layers must be *c*, to prevent NTh_6 octahedra sharing faces (cf. previous footnote). Once again this results in the sequence of (0001) anion layers being . . . *abcabc* . . .

The structure is shown in Fig. 2. There are now two different cation coordinations Th(2)N_7 (as LaO_7 in A-La_2O_3) and octahedral Th(1)N_6. Bond lengths are respectively 2.30$_9$ Å (3 ×), 2.47$_1$ Å [all to N(1), tetrahedrally coordinated by Th], and 2.90$_9$ Å (3 ×) and 2.53$_2$ Å (3 ×) [to N(2), octahedrally coordinated by Th]. These bond lengths result from a complex interplay of cation . . . cation and anion . . . anion repulsions but, generally, those to tetrahedrally coordinated N are shorter than those to octahedrally coordinated N.

The cation layers are, in this case, spaced at almost equal intervals, *d(h–h)* = 3.03$_8$ Å, *d(h–c)* = 3.04$_6$ Å. The axial ratio is a trifle low, *c/a* = 7.068 compared with the ideal $(9/2)\sqrt{8/3} = 3\sqrt{6}$ = 7.348 for perfect eutaxy. It is particularly noteworthy that the cation array is much more regular than the anion array: $d(\text{Th}\ldots\text{Th})$ varies from 3.77_3 to 3.87_5 Å, a range of $\pm 1.3\%$, whereas $d(\text{N}\ldots\text{N})$ varies from 2.93_6 to 3.87_5 Å, a range of $\pm 13.8\%$.

2.1.2.4. Ψ Phases in the Systems $Ln_2O_3 + MO_2$ (Ln = La, Pr, Nd, Sm, Gd; M = Th, Ce)

Sibieude[16, 17] has reported the occurrence of several ordered or partly ordered "Ψ" structures at high temperatures in these systems. The evidence is from X-ray powder-diffraction[16, 17] and electron microscopy[17]. In the light of our earlier discussion of B 1, C 1 and A-La_2O_3 it is not surprising that these ternaries are also intergrowths of {111} layers of the first two – or the last two.

In Th_3N_4, double layers of B 1 separate single layers of C 1. By contrast, in the present cases single layers of B 1 separate multiple layers of C 1. [Alternatively, single layers of A-type are intergrown with C 1 layers of varying thickness, which is not surprising since the sesquioxides are A-type (at high temperature) and the dioxides are fluorite type]. All the structures are hexagonal (or rhombohedral) with long *c* axes that are multiples of *c*/2 for A-type [= *d*(111) for C 1]. Multiplicities range from $n = 9$ to $n \approx 200$: X-ray diffraction resolved five *ordered* phases, Ψ_0, Ψ_1, Ψ_2, Ψ_3, Ψ_4, and three of their structures:

$\Psi_1 = 25\,\text{H}$;[5]

$\Psi_2 = 15\,\text{R}$; cation layer stacking $ch^4 = \ldots(\alpha\beta\alpha\beta\alpha\beta\gamma\beta\gamma\beta\gamma\alpha\gamma\alpha\gamma)\ldots$;

$\Psi_3 = 9\,\text{R}$; $ch^2 = \ldots(\alpha\beta\alpha\beta\gamma\beta\gamma\alpha\gamma)\ldots$;

$\Psi_4 = 12\,\text{R}$; $c^2h^2 = \ldots(\alpha\beta\alpha\beta\gamma\alpha\gamma\alpha\beta\gamma\beta\gamma)\ldots$.

Applying the previous "avoidance principle" allows one to deduce the layer sequences for the anions: it is invariably ... *abcabc* ... of course. The sequence of anion-occupied layers of tetrahedral (*t*) and octahedral (*o*) sites would then be:

A-La_2O_3 $= \ldots(to)\ldots$;

C 1 $= \ldots(t)\ldots$;

Ψ_2 $= \ldots(tot^2o)^3\ldots$;

Ψ_3 $= \ldots(t^2o)^3\ldots$;

Ψ_4 $= \ldots(t^3o)^3\ldots$

From this one can deduce an ideal stoichiometric ratio of C 1 to B 1, or MO_2 to Ln_2O_3. In general, the observed stoichiometry differs from the ideal value, indicating incomplete

5 The number is the value of *n*, and H = hexagonal, R = rhombohedral ($|n|_{\text{mod}\,3} = 0$)

occupancy of the anion sites. But this is not surprising for such high-temperature ($1200 \lesssim T \lesssim 1950\,°C$) structures. At these temperatures many fluorite-like structures have highly or completely disordered anion arrays – they are solid electrolytes[18)], and also commonly non-stoichiometric[19, 20)].

Electron-microscope studies of quenched, high-temperature $La_2O_3 + CeO_2$ preparations[17)] revealed the higher n values and, in most samples, imperfect ordering. Similar studies of arc-melted ThO_2 revealed stacking faults[21)], which are probably {111} layers of B 1 [= (0001) layers of A-type] also[11)] – a result of slight anion loss at the melting temperature[6]. [N.B. Contrary to Sibieude's assertion[16)], this is *not* the case in the homologous series of ordered structures Ln_nO_{2n-2} (Ln = Ce, Pr or Tb). In these the cation array is always very close to cubic eutaxy.]

2.1.2.5 ε-Hf_3N_2 and ζ-Hf_4N_3

These[22)] are constructed on a principle similar to that for **A-La_2O_3** and **Th_3N_4**, except that the "C 1 layers" are now devoid of anions, i.e. they are simply cations in cubic eutaxy. All the anions are in octahedral coordination, and a similar "avoidance rule" applies (c layers of cations except at B 1/A 1 boundaries where they are h). Hence, the stacking sequence of eutactic layers of cations is, h^2c in ε-Hf_3N_2 and h^2c^2 in ζ-Hf_4N_3; so that the complete sequences are (roman letters = anion layers, greek letters = cation layers)

$$\ldots \alpha c\beta(\alpha c\beta a\gamma\ \beta a\gamma b\alpha\ \gamma b\alpha c\beta)\alpha c\beta a\gamma \ldots = (h^2c)^3,$$

and

$$\ldots \alpha(\beta c\alpha b\gamma a\beta\ \gamma a\beta c\alpha b\gamma\ \alpha b\gamma a\beta c\alpha)\beta c\alpha b\gamma \ldots = (h^2c^2)^3$$

respectively.

By analogy with the foregoing structures, they may conveniently be regarded as $Hf\Box_2^t$ + 2 HfN and $Hf\Box_2^t$ + 3 HfN respectively (with HfN = B 1).

Finally, we may note that suboxides of metals like Ti and Zr also fit this picture: up to the oxygen content of M_2O they are anti-$Cd(OH)_2$ type, with incomplete occupancy of the octahedra in the B 1 layers for O/M < 0.5, complete for O/M = 0.5. The cation layer stacking is $h = \ldots \alpha\beta\alpha\beta \ldots$ in all cases.

2.2 *Perovskites*

The familiar cubic perovskite structure of ABO_3 has of course just one structural parameter, the unit cell edge a. This requires the ratio of the A–O to the B–O bond lengths to be equal to $\sqrt{2}$. When this condition cannot be met, the structure distorts in (one of) a number of well-documented ways[23, 24, 25)]. By far the largest of the families of derivative structures that arise when A is too small [$l(A\text{–}O)/l(B\text{–}O) < \sqrt{2}$] is that of the orthorhombic perovskites (**$GdFeO_3$** type) exemplified by the mineral perovskite ($CaTiO_3$) itself.

---

6 They could also be {111} layers of (anion-free) Th, cf. next section

The metal-atom arrangement in cubic perovskite is that of β-brass (**CuZn**), with oxygen in the B_2A_4 „octahedral" sites (more correctly in the B ... B edge common to four B_2A_2 tetrahedra). From the point of view of this paper, the distortion may be described as the oxygen moving from that B ... B edge to the interior of a B_2A_2 tetrahedron in such a way that B remains coordinated by six O atoms and A has six nearest-neighbour O atoms at the corners of a (more or less regular) trigonal prism. The derived structure has eleven free parameters, so that it is a large task to calculate them (but see refs.[24, 25] for a discussion in terms of *cation*-centred polyhedra).

A simpler (in the sense of fewer free parameters) structure derived from that of perovskite is that of calcite, $CCaO_3$. This may be obtained from the cubic structure by concerted rotations of CaO_6 octahedra about axes parallel to [111], so that the coordination of C is reduced from O_{12} to O_3[23, 25]. In terms of anion coordination, this corresponds to displacement of O atoms from the OCa_2C_4 "octahedra" (or Ca–O–Ca line) to an OCa_2C triangle. (There are also simultaneous, smaller relaxations of the cation positions.)

2.3 $BaSO_4$ (HO_2) Structure

Where this structure has been considered (and it usually is not), it has been found baffling. One author describes it, quite erroneously, as a **NaCl**-like array of barium ions and sulphate ions. Bragg and Claringbull[27] draw the structure showing the SO_4 tetrahedra and the 12 BaO bonds (to 7 different SO_4 groups!). Wyckoff[28] says only that the Ba coordination is indefinite, while elsewhere[29] he comments that "Although many complicated salts ... have fundamentally simple structures, there are numerous others which bear no obvious relation to the simpler arrangements. The sulphates of univalent (such as K_2SO_4 – see below) and of divalent (such as $BaSO_4$) metals are of this sort; their structures, departing far from atomic close packings, are dictated by the shapes and sizes of their ions. This inescapable dependence on the peculiarities of ionic shape dominates the structure of many salts having big and asymmetric ions; it makes the understanding of each a problem in itself": and this is for a structure, such as baryte, which is of very common occurrence – for sulphates, selenates, chromates, germanates, permanganates, perchlorates, fluoborates, etc. of large cations! [Wyckoff[28] lists about 30 compounds.] We may also note that Wells[24] seems not to consider metal sulphate structures at all!

The structure of $BaSO_4$[30] itself, and our interpretation of it, is shown in Fig. 3. The Ba + S array is that of a well-known alloy structure, **FeB**(B 15), shown in Fig. 4. Thus, in terms of the "complex ion" SO_4[7], the structure may be described as **FeB** type with SO_4 in place of **B** and Ba in place of **Fe** (as in the centre of Fig. 3) or, better, in terms of anion-centred polyhedra it is simply a BaS array of B 15 type with the oxygens inserted into SBa_3 tetrahedra. These anion-centred tetrahedra are irregular – as is to be expected, because the sizes of Ba and S are different.

At high temperatures some of these ABX_4 sulphates, perchlorates etc. transform to a cubic structure ($H0_5$ type)[31, 32]. While there is some doubt about the details (in particular

7 The difference between the Ba–O and the S–O bonds is surely only one of degree (and not of kind); the latter has a bond strength $s \approx 6/4 \approx 1\ 1/2$ compared with $s \approx 1/6$ for the former, a ratio of ~ 9

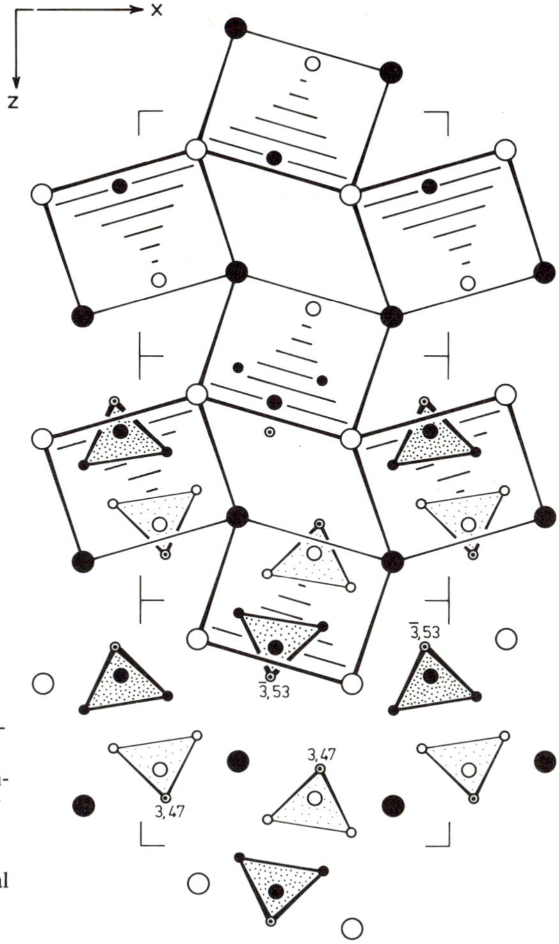

Fig. 3. The structure of $BaSO_4$ projected on (010). *Large, medium* and *small circles* denote Ba, S and O respectively. *Open/filled circles* at $y = 1/4/3/4$; *dotted circles* = O at heights given in units of $b/100$ ($y \approx 0,1/2$). *Top,* BaS array only (= **FeB**); *bottom,* usual depiction showing SO_4 tetrahedra and Ba atoms; *centre,* SO_4 tetrahedra and SBa_6 trigonal prisms. Unit cells indicated. Compare Fig. 4

Fig. 4. The structure of FeB projected on (010), space group setting *Pnma*. *Larger circles* are Fe, *smaller* are B; *open/filled* on $y/b = 1/4/3/4$. Two unit cells are outlined

the orientation of the BX_4 groups[33]) there appears to be no doubt that the cation array is **NaCl** (B 1) type (NaCl in $NaClO_4$; BaS in $BaSO_4$!). The large cation now has a more regular 12-coordination by anions (a truncated tetrahedron, slightly distorted if the BX_4 tetrahedra are tilted or librating). The coordination of the smaller cation by the larger one is still BA_6, but now an octahedron instead of a trigonal prism. A simple, martensitic mechanism is readily apparent[6]).

2.4 *Structures Derived From CrB (B 33)*

2.4.1 *β-CsBeF₃*

The structure of β-$CsBeF_3$[34]) is usually described as consisting of chains of corner-connected BeF_4 tetrahedra, with Cs atoms/ions interposed between them so that the Cs atoms are 8-coordinated by fluorine, but in a very irregular way. Figure 5 shows that, in fact, this Cs-centred coordination polyhedron is a rather irregular trigonal prism with two lopsided caps.

In contrast, Fig. 6 shows that if the fluorine is ignored the Cs and Be atoms form a perfectly regular array which is the **CrB** (B 33) structure (Fig. 7): Be-centred Cs_6 trigonal prisms share triangular faces to form columns parallel to *y*, and "square" faces to form

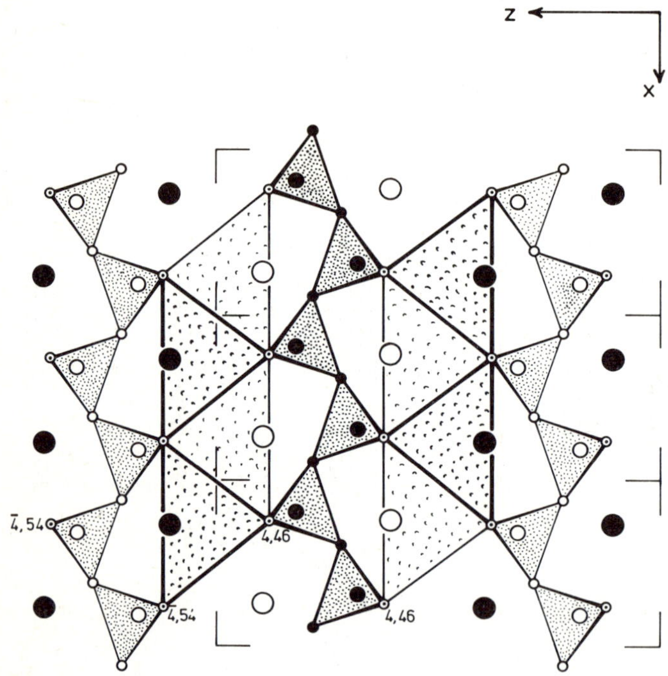

Fig. 5. The structure of $CsBeF_3$ projected on (010). *Large, medium* and *small circles* denote Cs, Be and F respectively. *Open/filled circles* at $y = 1/4 / 3/4$; *dotted circles* = F at heights given in units of *b*/100 ($y \approx 0,1/2$). Unit cells indicated. This is the conventional description with chains of BeF_4 tetrahedra and CsF_6 trigonal prisms

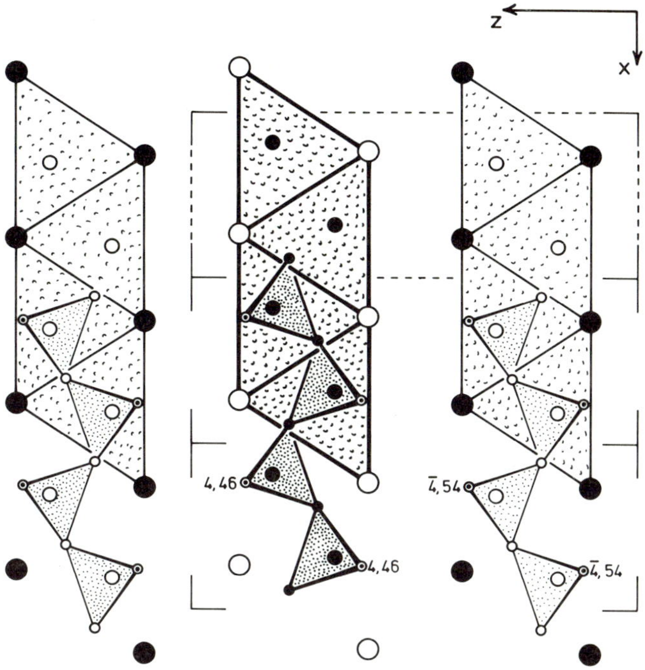

Fig. 6. $CsBeF_3$ projected as in Fig. 5. *Top,* CsBe array only (= **CrB**); *bottom,* usual depiction as BeF_4 tetrahedra plus Cs atoms; *centre,* BeF_4 tetrahedra and $BeCs_6$ trigonal prisms. Unit cells outlined

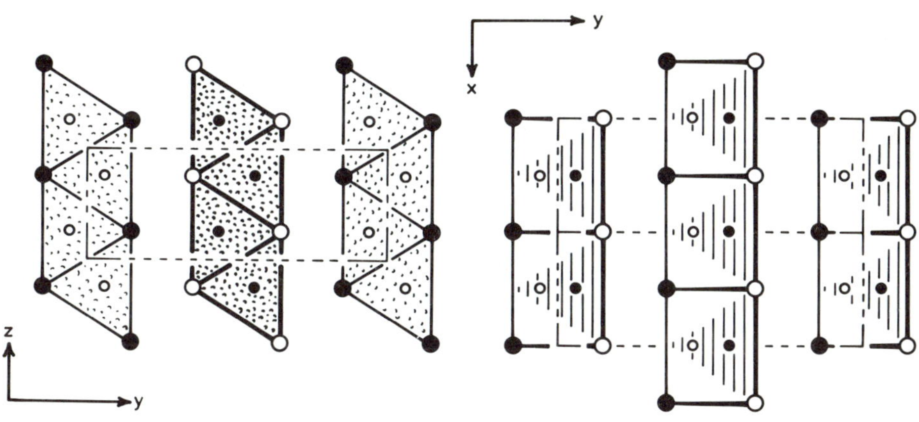

Fig. 7 a, b. The structure of CrB; *large circles* = Cr, *small circles* = B: (**a**) projected on (100), *open/closed circles* at $x = 0/1/2$; (**b**) projected on (001), *open/closed circles* at $x = 1/4/3/4$. Unit cells are outlined

walls in the (001) plane. Reasonably, the F atoms are all in tetrahedral interstices of the **CrB**-like array of CsBe: bridging atoms (in the BeF_3 chains) in Be_2Cs_2 tetrahedra (off-centre, and irregular for obvious reasons – Be and Cs being very different in size), and terminal F in Cs_3Be tetrahedra.

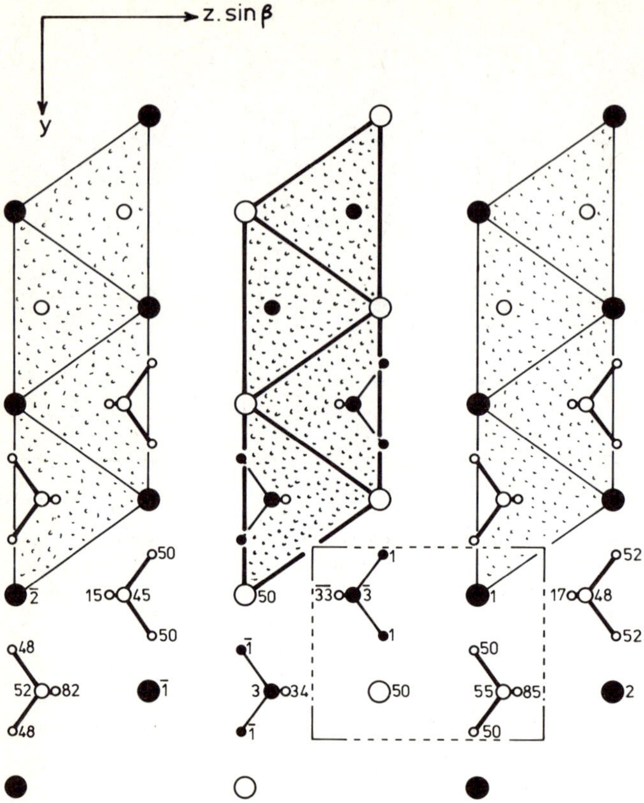

Fig. 8. The (room-temperature) structure of $KClO_3$ projected along [100] {*not* on (100)}, but atom heights (in units of *a*/100) measured above a plane normal to [100]. (The structure is almost orthorhombic.) *Large, medium* and *small circles* are K, Cl and O respectively. *Filled* and *open circles* are at $x \approx 0$ and 1/2 respectively (K, Cl and 2/3 of O: remaining O at $x/a \approx \pm 1/6, \pm 1/3$). A projection of one unit cell is outlined. *Top,* **CrB**-like KCl array; *bottom,* ClO_3 pyramids; *centre,* both pyramids and ClK_6 trigonal prisms

2.4.2 $KClO_3$ (GO_6)

This is rather similar to $CsBeF_3$: the structure of the room temperature form[35)] shown in Fig. 8 reveals the resemblance (cf. Fig. 6). The KCl part is again **CrB** type, but isolated ClO_3 groups occur instead of chains of BeF_4 tetrahedra. (This is, of course, a result of the "stereo-chemically active" lone pair of electrons completing the tetrahedral coordination of the chlorine, ClO_3E^-; it is really an ABX_4 structure, $KCl(O_3E)$. Note that the perchlorate, $KClO_4$, has the baryte structure considered earlier.) The structure is monoclinic ($P2_1/m$) rather than orthorhombic (*Pnma* for $CsBeF_3$ and CrB), but Fig. 9, a different projection of $KClO_3$, shows that the departure of the KCl array from orthorhombic symmetry is not very great. (The equivalent monoclinic angle for the truly orthorhombic structure would be 109° 16′ instead of the actual 108° 43′.)

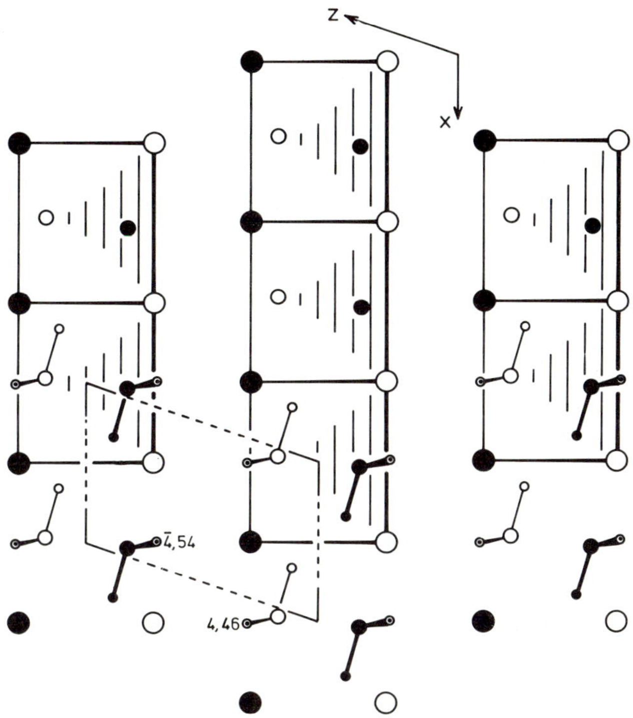

Fig. 9. Room-temperature $KClO_3$ projected on (010); cf. Fig. 8

Nevertheless, this slight departure from orthogonality is significant: at 250 °C $KClO_3$ transforms to a high-temperature, orthorhombic polymorph (space group *Pnma*), shown in Fig. 10 (cf. Fig. 9). The structure[36)] shown was determined at 280 °C. Changes in parameters are small, mainly slight reorientation of the ClO_3 pyramids. The equivalent lattice parameter changes are (low, *l;* high, *h;* in Å):

$$4.6535(a_l) \rightarrow 4.74(c_h), \quad \Delta = +1.86\% \;;$$

$$5.5841(b_l) \rightarrow 5.64(b_h), \quad \Delta = +1.00\% \;;$$

$$13.3573(^8) \rightarrow 13.80(a_h), \quad \Delta = +3.31\% \;.$$

And note that this range (an average of about 2% in linear terms) includes that for normal thermal expansion from 25 °C to 280 °C (~1/2%?).

Neither structure has previously been described simply. In both, the anions are inserted into ClK_3 tetrahedra (of two types in each structure) in a simple **CrB** array of KCl.

8 The axis in l equivalent to a in $h = \sqrt{a_l^2 + 4c_l^2 - 4a_lc_l\cos(180° - \beta_l)}$

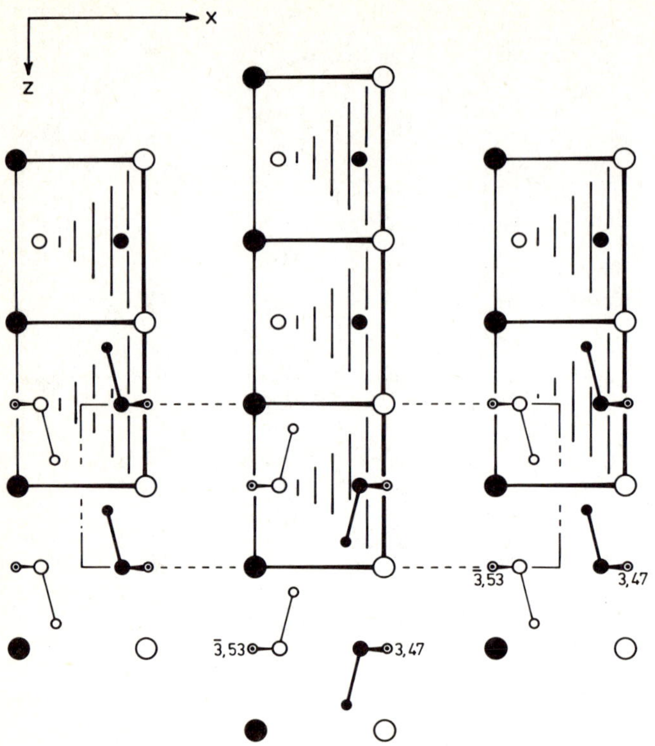

Fig. 10. High-temperature (280 °C) $KClO_3$ (orthorhombic) projected on (010). Compare Fig. 9

2.5 *Structures Based on the Ni_2Si (C37) or $PbCl_2$ (C23) Structure*

2.5.1 *β-K_2SO_4 ($H1_6$)*

This is another very common structure type with about 35 examples amongst metal sulphates, selenates, fluophosphates, chromates, vanadates, arsenates, fluo-borates and -beryllates, permanganates, halides such as Cs_2CuCl_4, and at least one titanate, Ba_2TiO_4, listed by Wyckoff[37]. Again, only SO_4 tetrahedra appear to have been recognised in the structure, Wyckoff commenting (in the case of Ba_2TiO_4) that "The coordination of the barium atoms is uncertain as is often the case, with Ba–O rising by small increments from 2.56 Å"[37].

The example shown in Fig. 11 is the structure of β-K_2SO_4 itself[38]. At the top the K_2S array is emphasised, and the structure becomes abundantly clear: S is in tricapped trigonal prisms of K, the combined array being that of the **$PbCl_2$** (C23) type – a type well represented by alloys such as Ca_2Sn, Ca_2Si, as well as large-cation dihalides (see below). As with $PbCl_2$ itself, there is a range of distances between the central atom (in this case S) and the ligands (in the present context K)[6]. In β-K_2SO_4 the range is from 3.297 to 4.164 Å. Using the shortest as unit length, they are:

Fig. 11. The structure of β-K_2SO_4 projected on (010). *Large, medium* and *small circles* = K, S and O respectively: *open/filled* at $y = 1/4\,/\,3/4$; *dotted circles* = O at heights given in units of $b/100$ ($y \approx 0$, 1/2). *Top,* C23 array of K_2S; *bottom,* usual depiction showing SO_4 tetrahedra and K atoms; *centre,* SO_4 tetrahedra in K_6 trigonal prisms. Unit cells outlined

for β-K_2SO_4 1.000(1 ×) 1.018(1 ×) 1.030(1 ×) 1.038(2 ×) 1.077(2 ×) and 1.263(2 ×), cf. for $PbCl_2$ 1.000(1 ×) 1.009(2 ×) 1.076(1 ×) 1.079(1 ×) 1.080(2 ×) and 1.274(2 ×). It is clear then that the coordination number in both cases is 9 = 6 (trigonal prism) + 3 (caps); that in each case it is really 7 + 2 (cf. bond lengths); and that not only is the topology the same but the geometry almost identical too[9]. β-K_2SO_4 is one of *many* compounds and structure types that may be accurately described as anion-stuffed C23 or C37.

There are three crystallographically distinct anions: O(1) is in a K_4S trigonal bipyramid with axial S, and l(K–O) ranging from 2.72 to 3.13 Å, l(S–O) = 1.46 Å; O(2) is in a K_5S octahedron with l(K–O) ranging from 2.73 to 3.19 Å, l(S–O) = 1.56 Å; O(3) is also in a K_5S octahedron, with l(K–O) from 2.74 to 3.09 Å, l(S–O) = 1.47 Å.

---

9 The ratio of the height to the mean edge length of the base of the trigonal prisms is also not untypical of the unfilled structure type, being 1.47; cf. 1.45 in C37 (**Co_2Si**), 1.18 in C23 (**$PbCl_2$**)

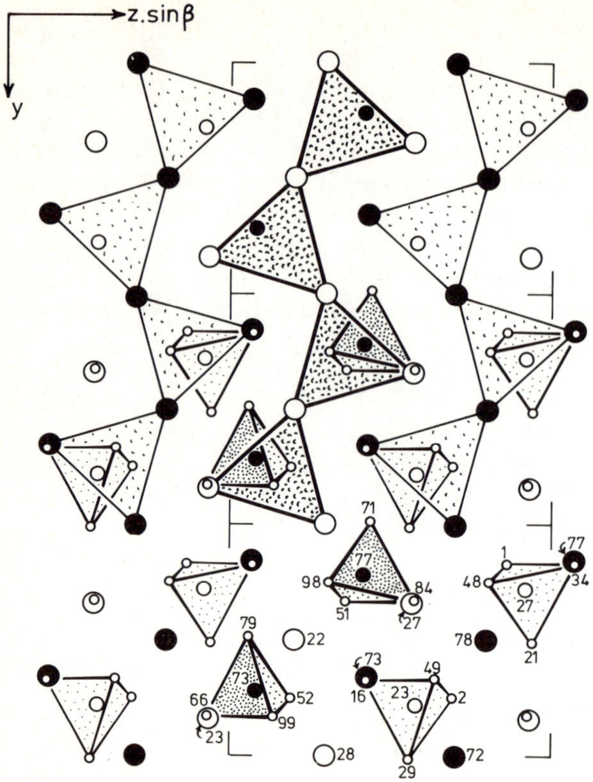

Fig. 12. The structure of β-Ca_2SiO_4 projected along [100] {not on (100)}. *Large, medium* and *small circles* = Ca, Si and O respectively. All atoms at heights indicated [in units of *a*/100 above (100) plane], Ca and Si at $x \approx 1/4$ und 3/4 respectively are *open* and *filled circles. Top,* Ca_2Si array of Si-centred tricapped trigonal prisms of Ca connected as in C23; *bottom,* the conventional drawing showing SiO_4 tetrahedra and Ca atoms; *centre* SiO_4 tetrahedra and Ca_6 trigonal prisms. Unit cells outlined

One might also point out that while the compound K_2S itself has the antifluorite structure (C1) under normal conditions, C1 often transforms to C23 under pressure, which may be expected to be the (not very) high-pressure form of K_2S. Thus (a situation we shall return to) β-K_2SO_4 may be regarded as an anion-stuffed, high-pressure K_2S polymorph.

2.5.2 β-Ca_2SiO_4, Larnite (Together With Some Discussion of Ca_2Si and Also of Ca and CaF_2)

This structure[39], shown in Fig. 12, closely resembles that of β-K_2SO_4 (Sect. 2.5.1). The main differences, and they are small, are *a*) a tilting of the BX_4 tetrahedra and *b*) a slight monoclinicity in the β-Ca_2SiO_4 unit cell (as with low-$KClO_3$, Sect. 2.4.2, due to slight tilting of the BA_2 sheets of trigonal prisms). But these are *minor* differences: the $SiCa_2$ array is again C23-like[10].

Three quarters of the oxygen atoms are tetrahedrally coordinated ($SiCa_3$), one quarter has $SiCa_4$ trigonal bipyramidal coordination. (Cf. K_2SO_4 above.)

This structure is particularly interesting in that it is another case in which the cation array in an ionic compound of complex structure is isostructural with the array of the same cations in the (unstuffed) parent alloy. Ca_2Si itself has the C23 structure; it is shown in Fig. 13. The relation to the upper part of Fig. 12 is striking. The symmetries and space groups are different – orthorhombic *Pnma* for the alloy, monoclinic $P2_1/n$ for the oxide – but the A_2B topologies are the same, and the parameter differences trivial. Even more striking perhaps is the relation between the unit-cell volumes: for β-Ca_2SiO_4 it is 343.9 Å^3, for Ca_2Si 331.2 Å^3. That is, if β-Ca_2SiO_4 were produced from Ca_2Si by simply inserting oxygen into the appropriate interstices, the changes would be a slight distortion (orthorhombic to monoclinic) and a volume increase of only 3.8% – approximately zero! (Indeed the trigonal prisms in β-Ca_2SiO_4 are slightly *smaller* than in Ca_2Si, but the prism sheets slightly further apart.)

The detailed comparison is as follows:

β-Ca_2SiO_4 (m)			Ca_2Si (o)	Δ (o → m)
a_m	= 5.502 Å	→	b_o = 4.799 Å	+ 14.65%
b_m	= 6.745	→	a_o = 7.667	– 12.03%
c_m	= 9.297		–	
β_m	= 94.59°	→	90°	
$c_m \sin\beta_m$	= 9.267 Å	→	c_o = 9.002	+ 2.94%
V_m	= 343.9 Å^3	→	V_o = 331.2 Å	+ 3.8%

The main change is an increase in $a_m \equiv b_o$ and a decrease in $b_m \equiv a_o$. This is an elongation of the $SiCa_6$ trigonal prism (and corresponding reduction in their cross-section). Such a change is to be expected because the number of Ca caps on the prisms – 3 in Ca_2Si – is "almost 4" in β-Ca_2SiO_4: compare Figs. 13 and 12. That is the Ca_2Si array in the latter is closer to C37 (**Co_2Si**) than to C23 (**$PbCl_2$**). This relation, and the effect of the number of caps on the relative height of the trigonal prism has been considered in detail elsewhere[6,25)].

This striking structural relationship between oxide and alloy is by no means unique (cf. Sect. 2.3, above). Although there is often a considerable volume increase on forming an inorganic structure from the corresponding alloy "by (notionally) inserting anions into the latter"[11], there are other cases like Ca_2Si → β-Ca_2SiO_4 where the volume change is close to zero. We may recall the very simple case of Ca → CaF_2. Both structures have Ca in cubic eutaxy; in the latter all the tetrahedral interstices are occupied by F ions, i.e. CaF_2 is produced from Ca metal by (notionally) inserting F into all its tetrahedral sites.

10 Also reported[40,41)] is another polymorph α'-Ca_2SiO_4 which is orthorhombic, the unit cell volume of 344.8 Å^3 being almost the same as that of β-Ca_2SiO_4, 343.9 Å^3. It seems likely[41)] that the α' polymorph is only a small deformation of β-Ca_2SiO_4. Contrast γ-Ca_2SiO_4 (Sect. 2.6.2) which is also orthorhombic, but with a much larger unit-cell volume, viz. 385.3 Å^3

11 Obvious examples being Co_2Si, Ni_2Si → Co_2SiO_4, Ni_2SiO_4: C37 → stuffed B8_b: cf. below

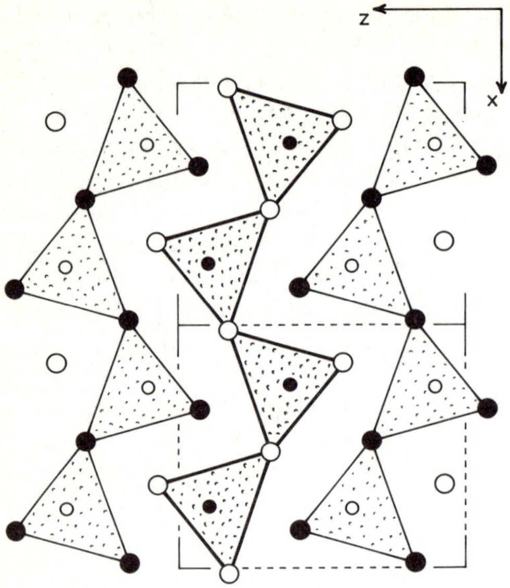

Fig. 13. The structure of Ca_2Si. *Large* and *medium circles* are Ca and Si respectively; *open* at $y = 1/4$, *filled* at $y = 3/4$. Compare Fig. 12, β-Ca_2SiO_4

The molar volumes are 43.622 and 40.759 Å^3 for Ca and CaF_2 respectively; i.e. Ca → CaF_2 is accompanied by a volume *decrease* of 6.6%! The f.c.c. unit cell edges are a = 5.588 Å for Ca[42] and 5.46295 Å for CaF_2[43) 12].

2.6 Structures Based on the Ni_2In ($B\,8_b$) Structure

2.6.1 High-Temperature K_2SO_4

This structure was determined at 800°[45]. It is shown in Fig. 14, which requires some explanation. The space group is hexagonal, $P6_3/mmc$, and in the figure it is projected on to the ($11\bar{2}0$) plane [(100) of the orthohexagonal unit cell whose axes are indicated on the drawing]. Clearly the sulphate groups are disordered between two orientations (though we suspect that this may be a result of twinning rather than a random distribution).

The K_2S array is topologically equivalent to that in β-K_2SO_4: columns of triangular-face-sharing SK_6 trigonal prisms united by edge-sharing into walls. But whereas these walls are puckered (zig-zag) in the β-form they are stretched (extended to the maximum extent in the $\mathbf{z}_o$ direction) in the high-temperature form. The latter array is the **Ni_2In** ($B\,8_b$) type – also hexagonal, space group $P6_3/mmc$. The relation between the C23 array of β-K_2SO_4 and this $B\,8_b$ array of the high-temperature form is rather obvious. It involves only a straightening of the sheets of trigonal prisms with increasing temperature. (The SO_4 orientation is very little altered.) Hence, a simple, diffusionless mechanism for the transformation is readily visualised, and leads one to expect considerable twinning on cooling.

12 As Lebedev[44], has pointed out the change for Mg → MgO is even greater, $\Delta V \approx -20\%$. (Mg and MgO both have eutactic metal atom arrays, hexagonal and cubic respectively)

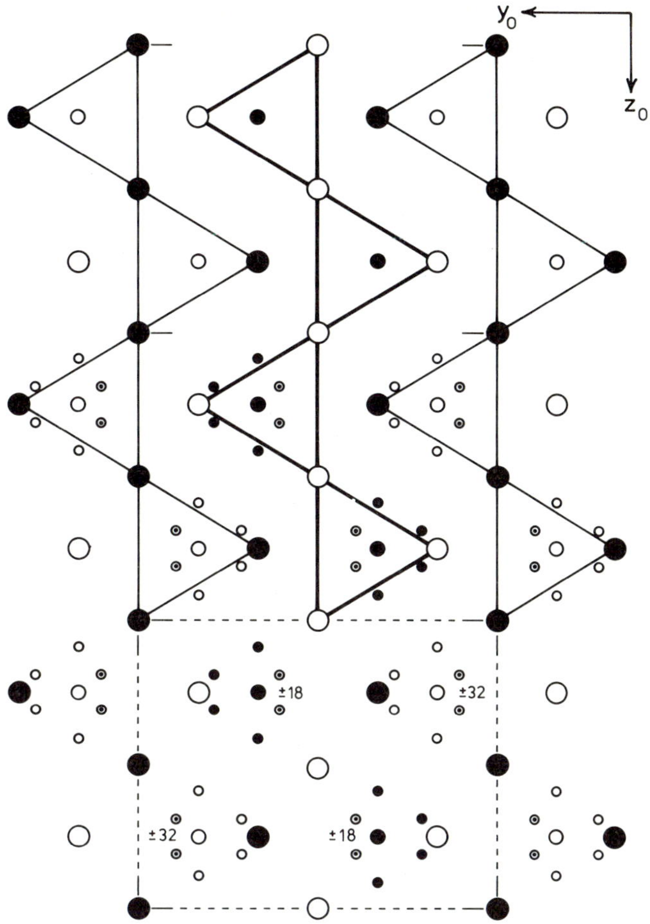

Fig. 14. High-temperature K_2SO_4 structure projected on $(11\bar{2}0)$ of hexagonal unit cell = (100) of orthohexagonal cell ($\mathbf{a}_o = \mathbf{a}_h + \mathbf{b}_h$, $\mathbf{b}_o = -\mathbf{a}_h + \mathbf{b}_h$, $\mathbf{c}_o = \mathbf{c}_h$). *Large, medium* and *small circles* = K, S and O; *open/filled* at x_o = 1/2 / 0; *dotted* at heights indicated in units of a_o/100. There are twice as many 0 sites as atoms because there are two (obvious) possibilities for orienting each SO_4 tetrahedron. *Top,* B 8_b- (Ni_2In-) like K_2S array. Compare Fig. 11 and Fig. 16.

The relation between the unit cells is as follows:

High-temperature K_2SO_4 (B 8_b, orthohexagonal)		β-K_2SO_4 (C 23, orthorhombic)	
$\mathbf{a}_o$	= 5.947 Å	**b**	= 5.763 Å
$\mathbf{b}_o$	= 10.301	**c**	= 10.071
$\mathbf{c}_o$	= 8.375	**a**	= 7.476
Volume	= 513.05 Å^3	Volume	= 433.90 Å^3

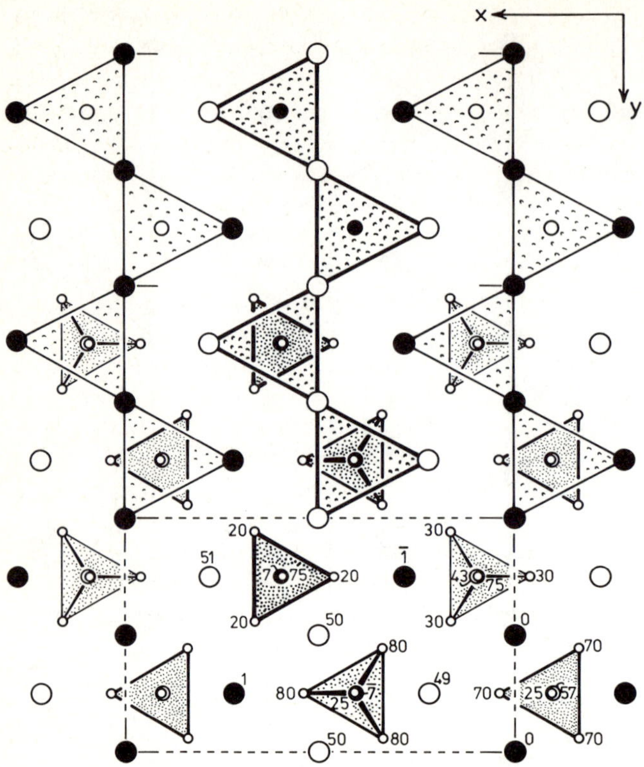

Fig. 15. The structure of γ-Ca_2SiO_4 projected on (001). *Large circles* are Ca, *medium circles* are Si, and *small circles* are O. Compare Figs. 12 and 16

As a result of the C23 → B 8_b transformation, the coordination number of S by K increases from ~6 + 3 to ~6 + 5: tricapped trigonal prisms to pentacapped trigonal prisms. Some aspects of this transformation in (unstuffed) alloys (such as MnCoGe) have been discussed elsewhere[6]. And a similar analysis of the same transformation in an oxide, $LiMgPO_4$, has recently appeared[46]. (Cf. also the next section.)

2.6.2 γ-Ca_2SiO_4

This polymorph is obtained by quenching the β-form (stable at higher temperature[40, 47]). Its structure[48] is shown in Fig. 15 (space group setting *Pnma*): it is olivine type, the Ca_2Si array being **Ni_2In** type. As in the case of K_2SO_4, it is interesting to compare the structure of the olivine-type polymorph with that of the β-form. (Though it may be noted that the β-form is a low-temperature polymorph of K_2SO_4, but a high-temperature – albeit metastable – polymorph of Ca_2SiO_4.) The cation arrangements (K_2S and Ca_2Si) are similar in the two systems (although in the β-forms the SO_4/SiO_4 tetrahedra are differently oriented): in both cases they are C23-like in the β-forms and B 8_b-like in the other polymorphs (high-temperature K_2SO_4 and γ-Ca_2SiO_4).

The β- → γ-Ca_2SiO_4 transformation is of some considerable interest, as it may be accompanied by "dusting" – a serious matter in cement, of which Ca_2SiO_4 is a constituent. There is doubt as to whether the transformation is reconstructive or displacive. In one paper[40)] it is decided that it is "semi-reconstructive"; in another[47)] no decision was reached. Comparison of the two structures involved is, in conventional terms involving emphasis on cation-centred coordination polyhedra, a complex task. It would appear to be much simpler in terms of the approach adopted in the present paper.

Changes in anion coordination (by cations) are readily seen: in the β-form there are four crystallographically distinct anion-centred coordination polyhedra; O(1), O(2) and O(3) are each in $SiCa_3$ tetrahedra, O(4) is in an $SiCa_4$ trigonal bipyramid: in the γ-form, of course, all anions are in $SiCa_3$ tetrahedra (cf. other olivine types above and below). But we suggest that these facts are not of great importance: the significant change is in the cation array – from C 23 in β to B 8_b in γ.

As already stated, this transformation is well known in (oxygen-free) metal alloys such as MnCoGe and MnNiGe, which are C 23 type at lower temperatures and B 8_b type at higher temperatures[49)]. There too the transformation results in the shattering of single crystals, even though it is diffusionless/displacive. This is attributed to the large change in molar volume (~ − 3.9% for C 23 → B 8_b in the case of MnCoGe)[49)]. Very careful and precise single-crystal X-ray diffraction studies of MnCoGe[50)] revealed that the thermal displacements of atoms in the B 8_b form corresponded exactly to those required to transform it to C 23, and *vice versa;* thus explaining its diffusionless character.

In the case of Ca_2SiO_4, the volume change for β → γ appears to be much larger (and in the opposite direction), + 12.0%[13]. But it seems entirely reasonable that there is a significant parallel between the two cases, oxide and alloy; and that the former can profitably be discussed in terms of the latter.

The principle cause of the volume change is an increase in the separation of the Ca_2Si sheets; 9.26_7 Å $= \mathbf{c}_\beta \sin \beta_\beta \rightarrow a_\gamma = 11.225$ Å. (The reverse transformation is never observed; β is always metastable.)

This (and high-temperature K_2SO_4, Sect. 2.6.1) is one of many compounds with the olivine structure – which we will now consider in detail.

2.6.3 Olivine ($H1_2/S1_2$)

Figure 16 shows the accurately determined structure of a natural olivine [Yosemite 103–481, $(Mg_{1.970}Fe_{0.024}Mn_{0.006})SiO_4 \approx Mg_2SiO_4$][51)]. The Mg_2Si array is **Ni_2In** (B 8_b) type. The sheets of $SiMg_6$ trigonal prisms are pushed further apart than in B 8_b, presumably to achieve appropriate **Si–O** and **Mg–O** bond lengths.

Elsewhere[6, 52)] the **Ni_2In** structure has been described in terms of unit cell twinning[53)]. The Ni atoms form a cubic eutactic array twinned (by reflection) on every second $(113)_{f.c.c.}$ plane. This is denoted as twinned c.c.p. ...*2,2,2,2,*... or, more briefly, (*2*). The twinning generates new interstices in the composition plane (on which the thin twin lamellae are joined) including trigonal prisms which, in Ni_2In, are occupied by In. Olivine is therefore an O-stuffed analogue: twinned c.c.p. Mg(*2*), with Si in the trigonal prisms. All the O atoms are then inserted into $SiMg_3$ tetrahedra (of three types).

13 There is some doubt about the exact value: reported unit cell parameters vary rather widely. This presumably reflects varying amounts and types of impurity, often added to stabilise the β-form

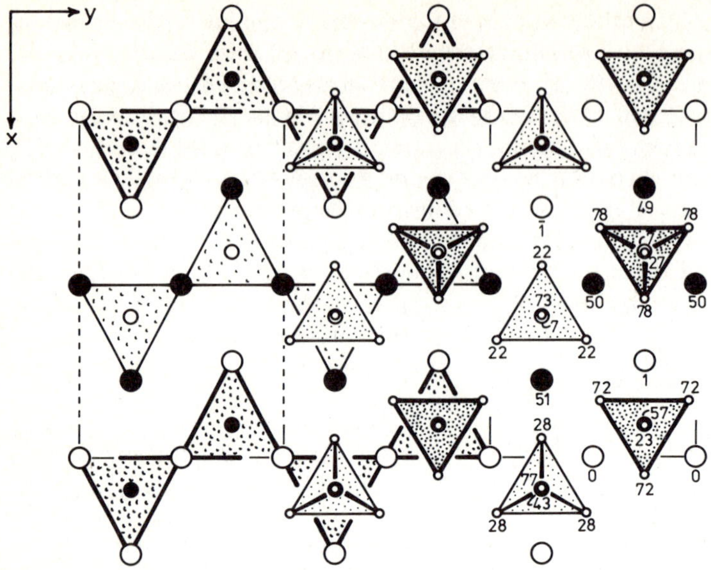

Fig. 16. The structure of olivine (Yosemite 103–481) projected on (001) (setting *Pnma*). *Large, medium* and *small circles* are ~Mg, Si and O; heights are in units of *c*/100. *Left,* ~Mg and Si only, emphasising B 8_b-like array of $SiMg_6$ trigonal prisms; *right,* SiO_4 tetrahedra and ~Mg atoms; *centre,* SiO_4 tetrahedra and ~Mg_6 trigonal prisms

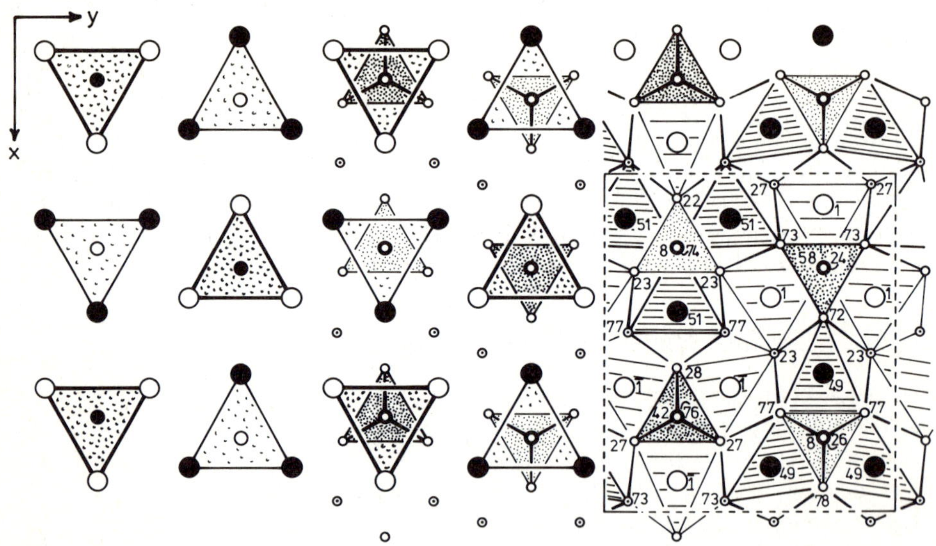

Fig. 17. The structure of norbergite, ~$Mg_2SiO_4 \cdot MgF_2$, projected on (001) (setting *Pnma*). *Large, medium* and *small circles* are respectively Mg, Si and O, and F (*dotted*); heights in units of *c*/100. *Left,* Mg and Si only, emphasising $SiMg_6$ trigonal prisms (as in **Re_3B**); *right,* the conventional depiction with SiO_4 tetrahedra and MgO_4F_2 octahedra; *centre,* SiO_4 tetrahedra in Mg_6 trigonal prisms, plus F atoms. (Note the regularity of the $SiMg_6$ prism outlines in contrast to the high distortion of the MgO_4F_2 octahedra – i.e. of the "h.c.p." array of anions.)

A virtue of the unit-cell-twinning description is that it becomes easy to see how additional, related structures can be generated to form "homologous series" – simply by varying the twinning interval, the width of the twin lamellae. Thus, the narrowest possible lamellae, i.e. twinned c.c.p. (*1*) generate the **Cr** array of the B 33, **CrB** structure (**B** in the trigonal prisms) that we used above (Sect. 2.4). Similarly, widening the twins to twinned c.c.p. (*3*) generates the **Re** array of the $\mathbf{Re_3B}$ structure type (**B** again in all the trigonal prisms) – which brings us to our next structure type.

2.7 Norbergite and the Humite Series (SO_7)

2.7.1 Norbergite, $\sim Mg_2SiO_4 \cdot MgF_2$

This structure is shown in Fig. 17, the data for which come from a natural specimen[54], $(Mg_{1.993}\ Fe_{0.002}\ Ca_{0.006}\ Zn_{0.001})\ SiO_4Mg_{0.989}\ Ti_{0.011}\ (F_{1.805}[OH]_{0.173}O_{0.022}) \approx Mg_2SiO_4 \cdot Mg(OH,F)_2$. It is orthorhombic (also space group *Pnma*), and the $\sim Mg_3Si$ array is that of the $\mathbf{Re_3B}$ (anti-$\mathbf{PuBr_3}$) type. (Re_3B itself is also orthorhombic, but space group *Cmcm.*) The Mg atoms form the twinned c.c.p. (*3*) array, with Si inserted into the trigonal prisms. The Si atoms are, of course, at the centres of O_4 tetrahedra. These oxygens are, as in olivine, at the centres of three types of $SiMg_3$ tetrahedra. Additional anions (largely monovalent F or OH) are in triangular Mg_3 coordination in the (010) bands between the trigonal prisms, i.e. they are the anions not bonded to Si.

Both olivine and norbergite (and the other members of the series, to be discussed next) are conventionally described as "hexagonal close-packed anion arrays with Mg in octahedral interstices and Si in tetrahedral interstices". The right-hand part of Fig. 17 depicts this conventional description by showing the cation-centred polyhedra. The distorted faces of the octahedra (isosceles rather than equilateral) indicate that the h.c.p. anion array description is a rough approximation only: the actual array is considerably deformed from h.c.p. (hexagonal eutaxy). This is also true of olivine. In striking contrast, the trigonal prisms of cations are rather regular in both structures. Hence the description of these two structures as anion-stuffed, regular cation arrays is rather more plausible than the conventional one of cation-stuffed h.c.p. arrays of anions.

2.7.2 Chondrodite, Humite and Clinohumite

The unit cell twinning description is easily extended to include asymmetric twinning, i.e. intergrowths of (*2*) and (*3*) such as ...*2,3,2,3,*..., ...*2,2,3,2,2,3*..., ...*2,2,2,3,2,2,2,3,*... etc., or (*2,3*), (2^2,*3*), (2^3,*3*) etc. Figures 18–20 show that these three examples elegantly describe the structures of the remaining three members of the chondrodite (or humite) series: chondrodite itself, humite itself and clinohumite, respectively. (And note that the "h.c.p." anion array is again badly distorted in *each* of these structures, but that the cation array is much more regular.) Because the twinning is asymmetric, chondrodite and clinohumite are monoclinic ($P2_1/b$). Humite is again orthorhombic (*Pnma*) because the crystallographic repeat must contain an even number of twin planes: it is therefore $(2^2,3)^2$ rather than (the simple layer-width repeat) (2^2,*3*). The drawings are also of accurately refined structures of mineral specimens:

chondrodite, $Mg_{4.95}\ Fe_{0.05}\ Si_2O_8\ F_{1.3}\ (OH)_{0.7} \approx 2\,Mg_2SiO_4 \cdot Mg(OH,F)_2$[55];
humite, $Mg_{6.6}\ Fe_{0.4}\ Si_3O_{12}F(OH) \approx 3\,Mg_2SiO_4 \cdot Mg(OH,F)_2$[56];
clinohumite, $Mg_{8.42}\ Fe_{0.50}\ Mn_{0.06}\ Ti_{0.02}\ Ca_{0.001}\ Si_4O_{16}\ F_{1.04}\ (OH)_{0.93}\ O_{0.03} \approx 4\,Mg_2SiO_4 \cdot Mg(OH,F)_2$[57].

As with norbergite and olivine, in each case the oxygen atoms (bonded to Si) are in "$SiMg_3$" tetrahedra and (as in norbergite) the monovalent anions (OH,F) are in Mg_3 triangular coordination between the trigonal prism bands, i.e. near the centres of the *3* twin lamellae. These $\mathbf{Re_3B}$ bands regularly interrupt the continuity of the $\mathbf{Ni_2In}$ sheets, which in olivine are continuous.

2.7.3 General Remarks on the Chondrodite/Humite Series

The original description on this SO_7 family of structures – $n Mg_2SiO_4 \cdot Mg(OH,F)_2$ with $n = \infty$ for olivine, 1 for norbergite, 2 for chondrodite, 3 for humite and 4 for clinohumite – was as intergrowths of unit lamellae of $Mg(OH,F)_2$(sellaite) with lamellae of olivine, the latter having different thicknesses in the different structures – see for example Bragg and Claringbull[27]. But objections can and have been raised against this description: *a)*

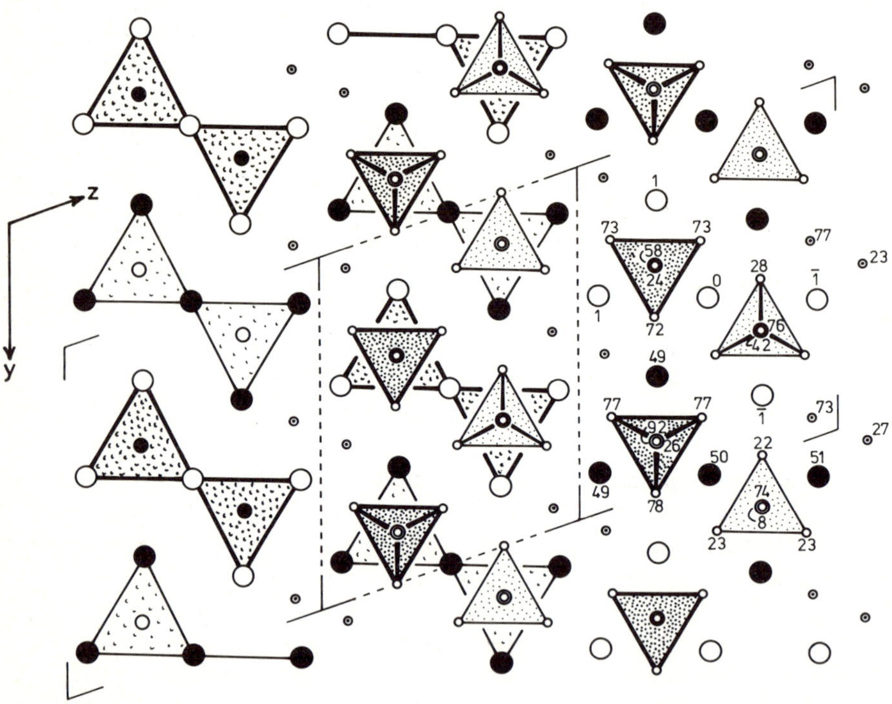

Fig. 18. The structure of chondrodite, $\sim 2\,Mg_2SiO_4 \cdot Mg(OH,F)_2$, projected on (100) (non-standard setting of $P2_1/b$). *Large, medium* and *small circles* are, respectively, "Mg" and Si (*open* at $x \approx 0$, *filled* at $x \approx 1/2$) and O (*dotted* = OH, F). *Left,* Mg and Si only, emphasising $SiMg_6$ trigonal prisms; *right,* SiO_4 tetrahedra plus Mg and (OH, F) atoms (MgX_6 octahedra not outlined); *centre,* SiO_4 tetrahedra and $SiMg_6$ prisms and (OH, F) atoms. Compare Figs. 16 and 17, and not that the MgX_6 octahedra (and therefore the "h.c.p." anion array) are grossly distorted

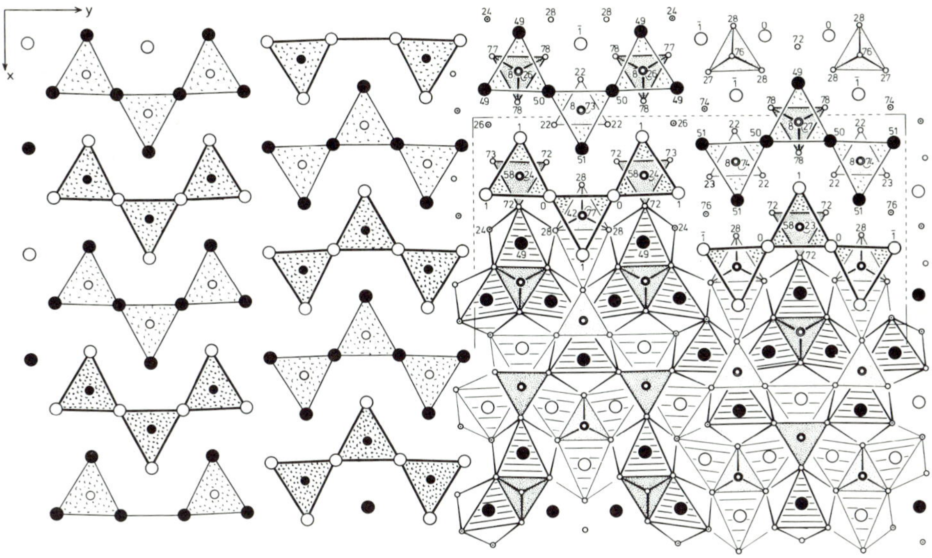

Fig. 19. The structure of humite, $\sim 3\,Mg_2SiO_4 \cdot Mg(OH, F)_2$, projected on (001) (*Pnma*). Atoms represented by *circles* as in Figs. 16–18. *Left,* $SiMg_6$ trigonal prisms only; *right, bottom,* conventional depiction as SiO_4 tetrahedra and MgX_6 octahedra; *top,* SiO_4 tetrahedra and Mg_6 trigonal prisms and (OH, F) atoms. Again note the distortion of the MgX_6 octahedra, and "h.c.p." anion array

the unit lamellae of sellaite are extremely thin (~1.56 Å), i.e. there is no real sellaite portion [with $Mg(OH,F)_6$ octahedra]; *b)* it is cumbersome – compare Bragg and Claringbull's AB · AB · AB; ALBR · ALBR; BRA · BRA; BABRABAL · BABRABAL and ABRAB · ABRAB for olivine, norbergite, chrondrodite, humite and clinohumite with our twinned c.c.p. *2; 3; 2,3;* 2^2,*3* and 2^3,*3*! [Note also that the number of *2* layers between nearest *3* layers is $(n - 1)$, where n is the integer in the general formula for the series.]

2.7.4 Leucophoenicite and Related Structures

The cation arrays in the humite series are intergrowths of $\mathbf{Ni_2In}$ with unit lamellae of $\mathbf{Re_3B}$. A similar series can be generated in which the cation arrays are intergrowths of $\mathbf{Ni_2In}$ with unit lamellae of **CrB**[58]. In the case of silicates, the only known example is leucophoenicite, a dimorph of manganhumite $3\,Mn_2SiO_4 \cdot Mn(OH)_2$[14].

The silicon positions in the *face-sharing* pairs of trigonal prisms (in the **CrB,** twinned c.c.p. *1* lamellae) are only half-occupied. They are at the centres of a pair of edge-sharing

14 Since this was written it has been suggested that the new mineral jerrygibbsite, $Mn_9(SiO_4)_4(OH)_2$, may be a second silicate member of the series[59]. This now appears to be confirmed: according to the structure determination of T. Kato (private communication, September 1983) jerrygibbsite has a $(1,2^4)^2$ cation array; cf. sonolite (Mn-clinohumite) = $(3,2^3)$. The relation between these two polymorphs is the same as that between leucophoenicite $(1,2^3)$ and manganhumite $(3,2^2)^2$ except that the Si atoms in the face-sharing Mn_6 trigonal prisms are ordered in jerrygibbsite but disordered (half-occupancy of both sites) in leucophoenicite

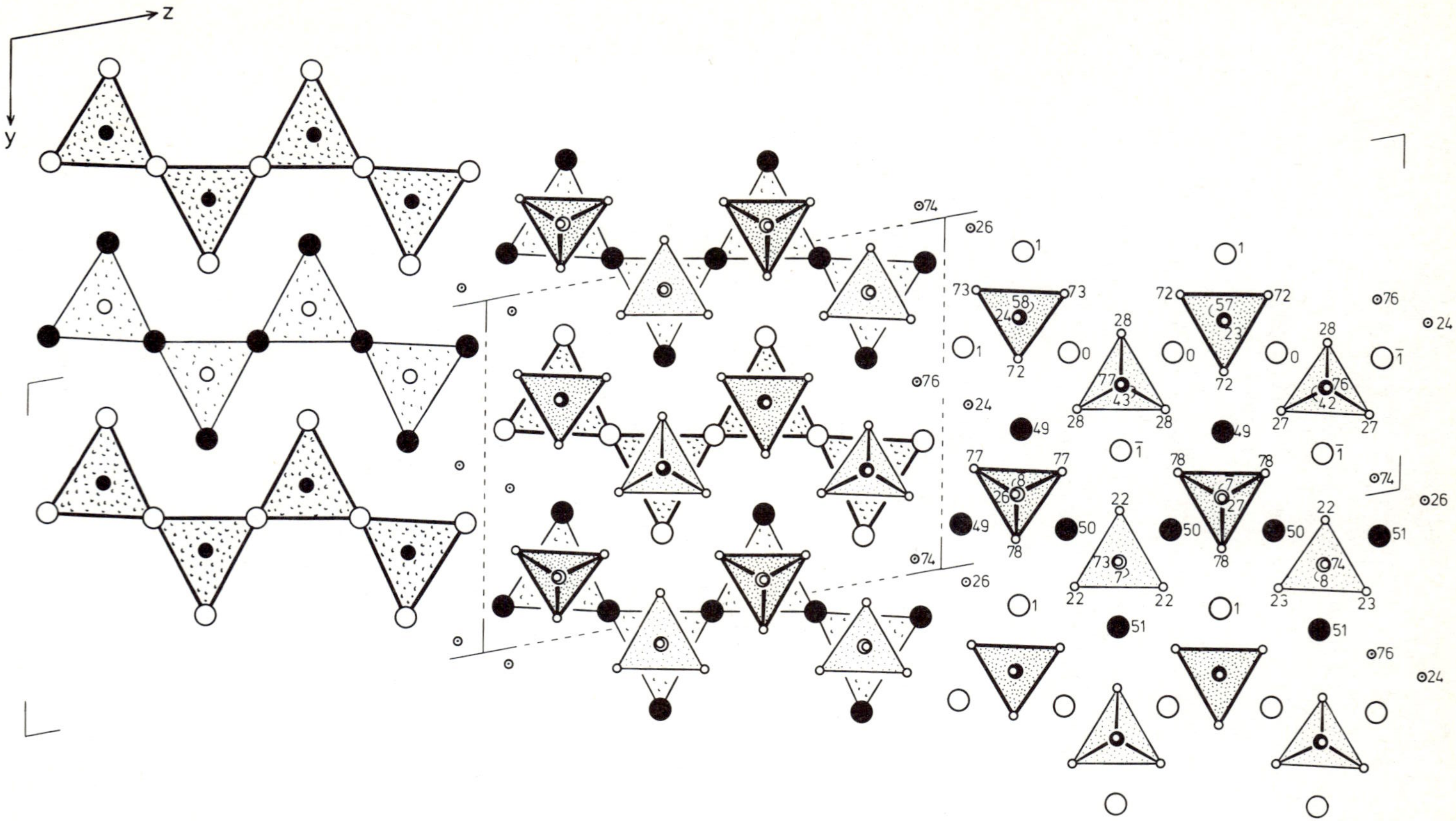

Fig. 20. The structure of clinohumite, $\sim 4\,Mg_2SiO_4 \cdot Mg(OH, F)_2$, projected on (100) (non-standard setting $P2_1/b$). *Circles* as in previous Figs. 16–19 [except that (OH, F) is now ($OH_{0.93}$, $F_{1.04}$, $O_{0.03}$)]. *Left,* Si and Mg atoms only, emphasising $SiMg_6$ trigonal prisms; *right,* SiO_4 tetrahedra and Mg atoms and (OH, F) atoms (*dotted circles*); *centre,* SiO_4 tetrahedra and Mg_6 prisms and (OH, F) atoms. Again the distortion of the MgX_6 octahedra is clear, contrasting with the relative regularity of the Mg_6 prisms

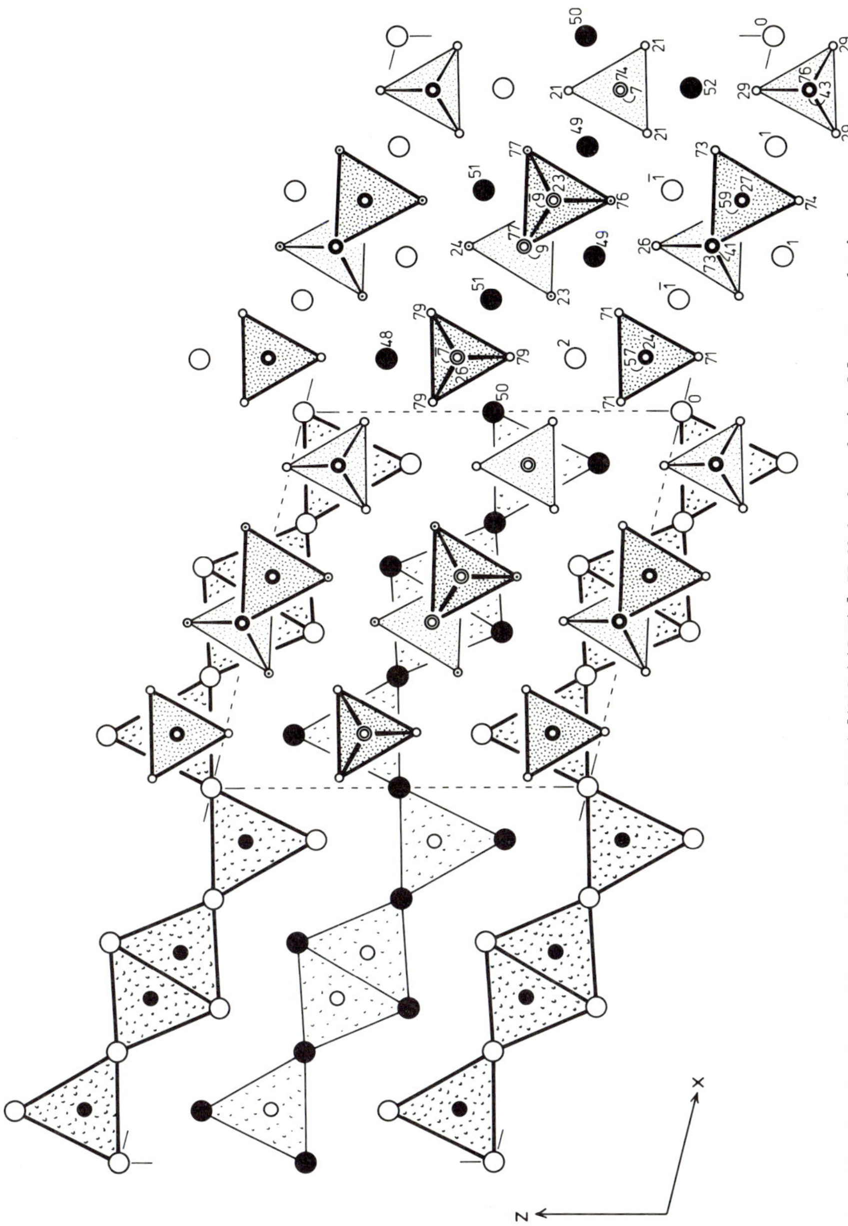

Fig. 21. The structure of leucophoenicite $Mn_7(SiO_4)_2[(SiO_4)(OH)_2]$. Half the face-sharing Mn_6 trgonal prisms are said to be occupied by Si in a statistical manner; the cation array twin formula is $(1,2^3)$. (N.B. Left-handed set of axes.)

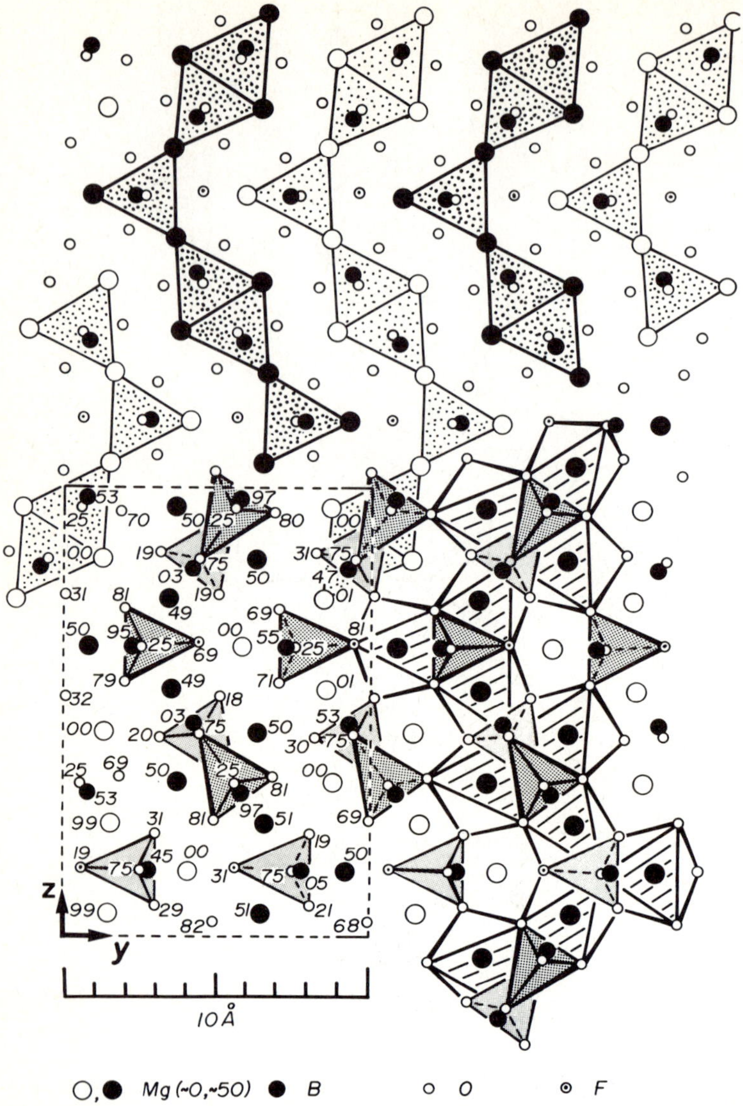

Fig. 22. The structure of $Mg_5(BO_3)_3F$. Its cation array is twinned c.c.p. ...*2,(1,2,2,1,2,2),1*... = $(1,2^2)^2$, an intergrowth of **CrB** and $\mathbf{NiIn_2}$. The B atoms are in triangular coordination, but are shown as BO_4 tetrahedra to emphasise the relation to leucophoenicite

tetrahedra of oxygen, O_6, and too close for both to be occupied simultaneously. The Si-free tetrahedron contains two hydrogen atoms, i.e. the group is $SiO_4(OH)_2$. The structure is shown in Fig. 21. The Mn array is twinned c.c.p. $(1,2^3)$.

No other silicates with arrays $(1,2^n)$ are known[15], except as defects in leucophoenicite crystals[60)] but other compounds are known. The cation array in magnesium fluorobo-

15 See previous footnote

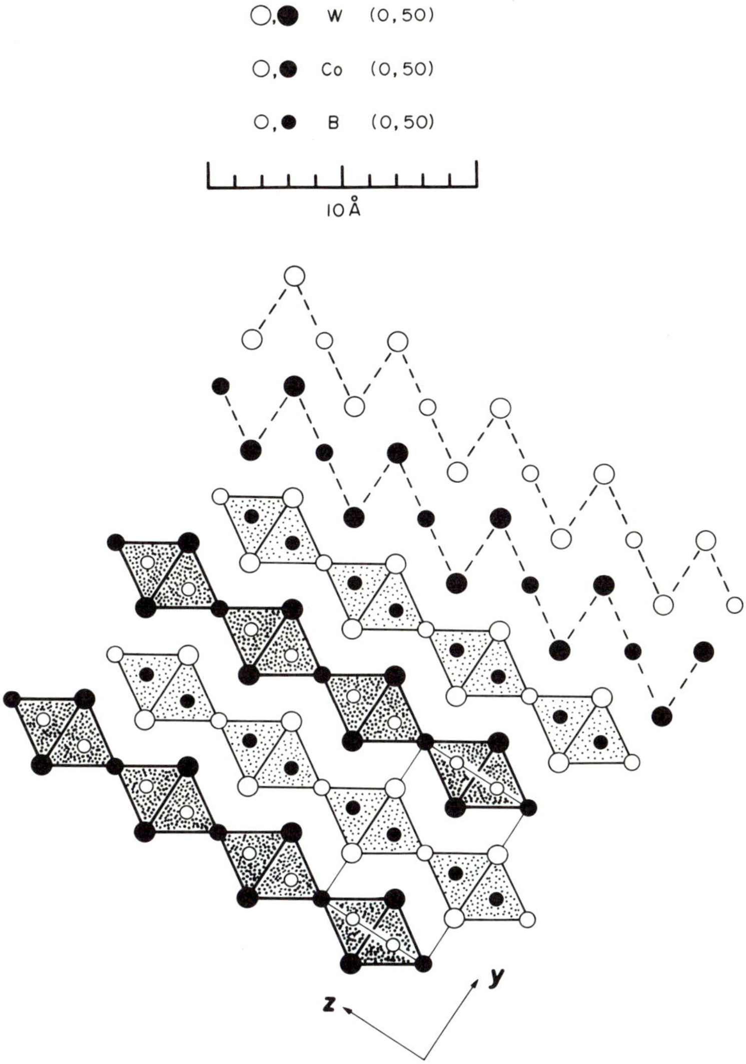

Fig. 23. The $\mathbf{W_2CoB_2}$ structure. It is twinned c.c.p. W_2Co ... *(1,2)*... forming face- and edge-sharing, boron-centred, trigonal prisms

rate[61]), $Mg_5(BO_3)_3F = 2\,Mg_2BX_4 \cdot MgBX_2$, Fig. 22 is twinned c.c.p. $(1,2^2)^2$. In it the B atoms are in Mg_6 trigonal prisms and are off centre in the O_4 tetrahedra, so much so that they are virtually in O_3 triangles. [This may be regarded as being due to strong B...B repulsion, of the same sort (but weaker than that) which precludes both tetrahedra being simultaneously occupied by Si in leucophoenicite.]

The twinned c.c.p. *(1,2)* cation array is known as an alloy – the $\mathbf{W_2CoB_2}$ type, Fig. 23 – as well as, for example, in the structure of kotoite, $Mg_3(BO_3)_2 = Mg_2BO_4 \cdot MgBO_2$, shown in Fig. 24, jimboite, $Mn_3(BO_3)_2$, and cadmium germanate, $Cd_3GeO_4(OH)_2$.

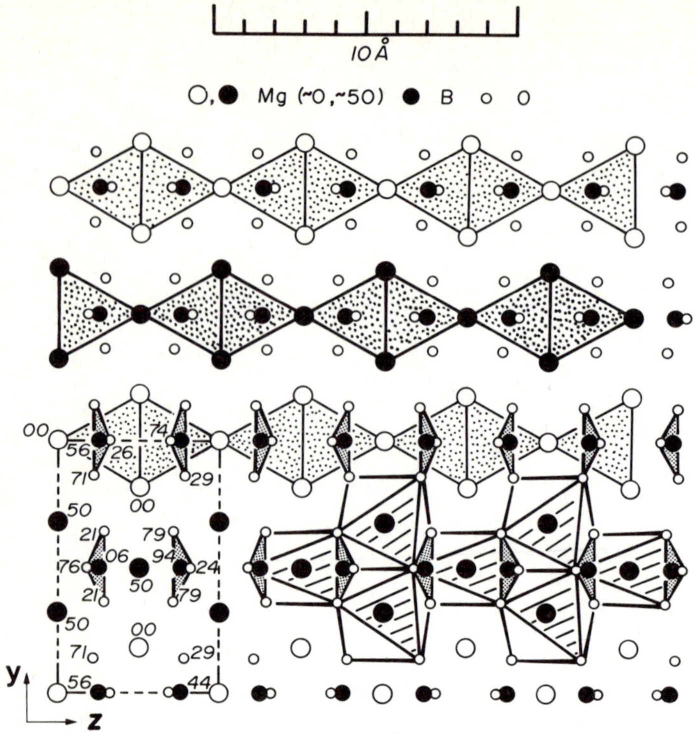

Fig. 24. Structure of kotoite, $Mg_3(BO_3)_2$. It is an anion-stuffed version of the $\mathbf{W_2CoB_2}$ alloy type with twinned c.c.p. Mg ... *(1,2)*.... (The B atoms are displaced from the centres of the trigonal prisms.)

Anion-stuffed, twinned c.c.p. *1* (**CrB**) cation arrays were considered in Sect. 2.4; those of twinned c.c.p. *2* ($\mathbf{Ni_2In}$) in Sect. 2.6. But an additional example of the latter is depicted in Fig. 25: cadmium chondrodite, $2\,Cd_2SiO_4 \cdot Cd(OH)_2$, in which the anion array is extremely distorted (from h.c.p.) although the cation array remains regular, emphasising even more strongly the point made earlier: that the "h.c.p. anion array" description is not entirely satisfactory for this family of structures.

2.8 The Spinel ($H1_1$) Structure

We consider a general spinel AB_2X_4. It has symmetry $Fd3m$ with A atoms in 8(*a*) 0,0,0 etc., B atoms in 16(*d*) 5/8,5/8,5/8 etc. and X atoms in 32(*e*) *u,u,u* etc. (with $u \approx 3/8$). The conventional description is that of an anion (X) array in cubic eutaxy (c.c.p.) which would be metrically perfect with $u = 3/8$. B cations are in one half of the octahedral interstices and A cations in one eighth of the tetrahedral interstices. Ever since the structure was first determined it has been puzzling that the larger ion (Mg in $MgAl_2O_4$) fills the "smaller" interstice (although of course the departure from ideal close packing, so that $u > 3/8$, suitably enlarges the tetrahedral interstice) – an observation that makes

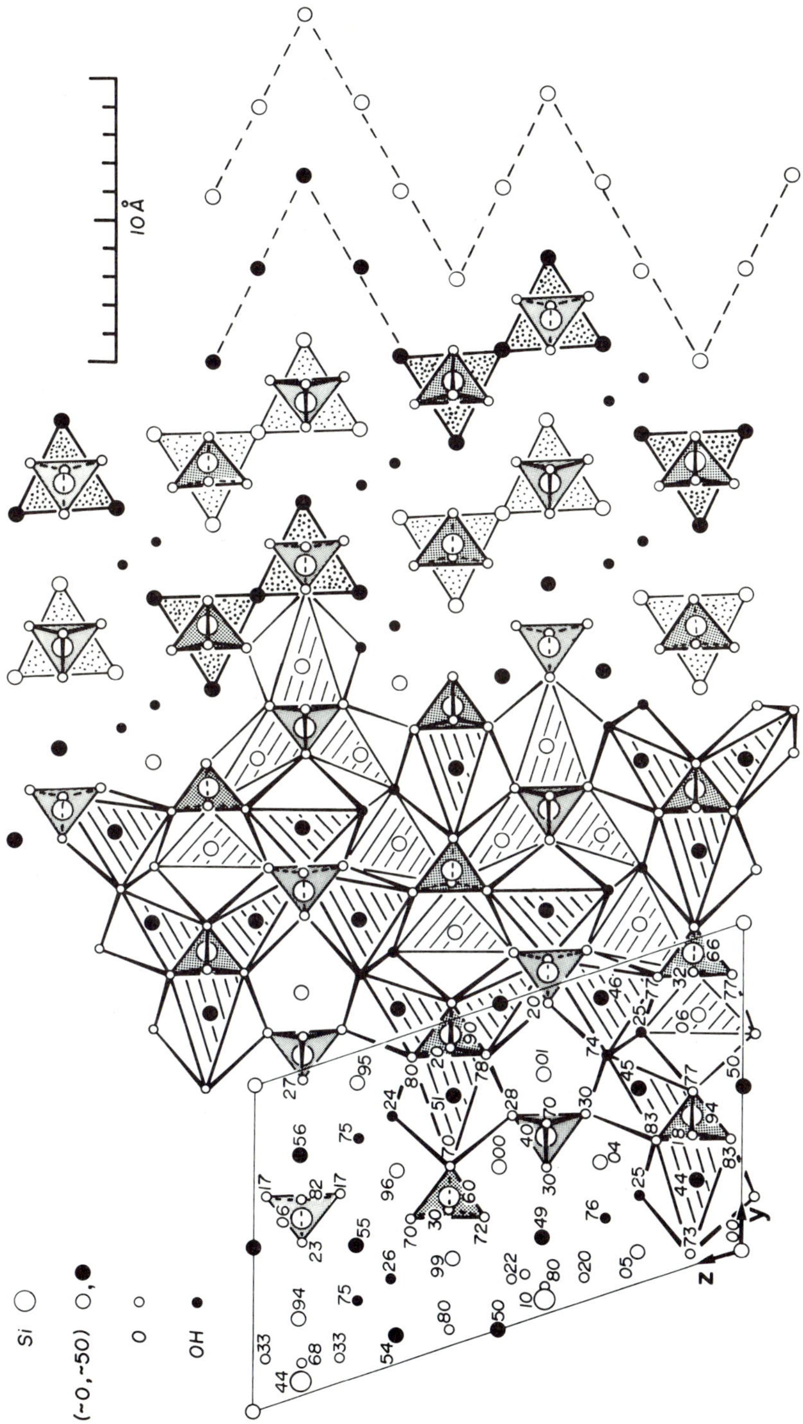

Fig. 25. Cadmium chondrodite, $2\,Cd_2SiO_4 \cdot Cd(OH)_2$, cation array twin formula (2,3). Both the conventional cation-centred polyhedra of anions and the intergrowth of **Ni_2In** (2) and **Re_3B** (3) structures are shown

nonsense of "radius-ratio" rules[16]. There has been endless speculation in the literature about the reasons for the structure adopting a certain value of the parameter u. As we shall see (in Sect. 3.2) this is virtually a non-problem as the value of u can be calculated for any normal spinel in an elementary way.

2.8.1 Geometrical Description of the Normal Spinel Structure

The AB_2 array is that of the cubic Laves phase, **$MgCu_2$**(C 15) type, with no undetermined parameters (except for the f.c.c. unit cell edge). It may be described in many ways (which are not very readily visualised because there is no very short axis along which the structure can be projected). In terms of polyhedra it is a space-filling array of empty tetrahedra (**Cu_4**) and filled, truncated tetrahedra (**$MgCu_{12}$**). The former are corner-connected in a C 9- ("ideal high-cristobalite"-) like array which is, in fact, exactly one half of a c.c.p. array of *Cu* atoms. The other half of a c.c.p. array would be identical and interpenetrating the first (as in Cu metal; and cf. Sect. 2.1.1.3, pyrochlore) but, in C 15, each of the **Cu_4** tetrahedra in this second array is replaced by one **Mg** atom at the position of the centre of the missing tetrahedron. The **$MgCu_{12}$** truncated tetrahedra thus formed share their edges and hexagonal faces (the **Mg** atoms therefore forming a diamond-like array). Figure 26 attempts to show this description of the C 15 structure: it is the $SiFe_2$ array in the cubic, spinel form of γ-Fe_2SiO_4[62] (which is produced from the normal olivine form by high pressure). In spinel itself, $MgAl_2O_4$, the Al atoms occupy the **Cu** sites of **$MgCu_2$,** the Mg atoms occupy the Mg sites and are tetrahedrally coordinated by oxygen. Thus, for the isostructural γ-Fe_2SiO_4, we get the corresponding Fig. 27 (in which, for clarity, only some Fe_{12} truncated tetrahedra and SiO_4 tetrahedra are drawn). And, in another projection [on (110) of the f.c.c. unit cell ≡ (100) of the equivalent, but smaller body-centred tetragonal unit cell] the same structural elements are emphasised in Fig. 28. Figure 29, equivalent to Fig. 28, shows the SiO_4 tetrahedra and the *C* 9-like network of (empty) corner-connected Fe_4 tetrahedra; this perhaps being the clearest representation of the structure.

Thus, instead of describing γ-Fe_2SiO_4 as C 15 **$MgCu_2$** with Fe in place of **Cu** and SiO_4 in place of **Mg,** it is best (in the present context) to regard it as a C 15 structure $SiFe_2$. Anions are then inserted into appropriate interstices all of which are $SiFe_3$ tetrahedra ($MgAl_3$ in $MgAl_2O_4$) and crystallographically equivalent. C 15 is a "topologically close-packed structure", in which space is divided into tetrahedra of three types: per AB_2 unit there are four AB_3 tetrahedra, one B_4 tetrahedron, and 12 A_2B_2 tetrahedra. In spinel only the four AB_3 tetrahedra are occupied by anions. This is depicted simply and incompletely in Fig. 30 (cf. Figs. 26 and 27). The tetrahedra are irregular, having an equilateral triangular base B_3, but isosceles AB_2 faces because $d(A \ldots B) > d(B \ldots B)$. The OAB_3 tetrahedra are corner-connected into groups of four by common A atoms, but edge-connected into groups of four by common B atoms. Thus, an alternative basic "building block" of the structure is a group of four OAB_3 tetrahedra sharing B . . . B edges, i.e. an empty B_4 tetrahedron (in the C 9-like array) capped on each face by an A atom to form a "stella quadrangula" of five tetrahedra, or a central B_4 tetrahedron sharing each face with

16 Some spinel-type compounds have $u < 3/8$ but they are high-pressure polymorphs, metastable at 1 atm pressure with respect to the olivine type (cf. below)

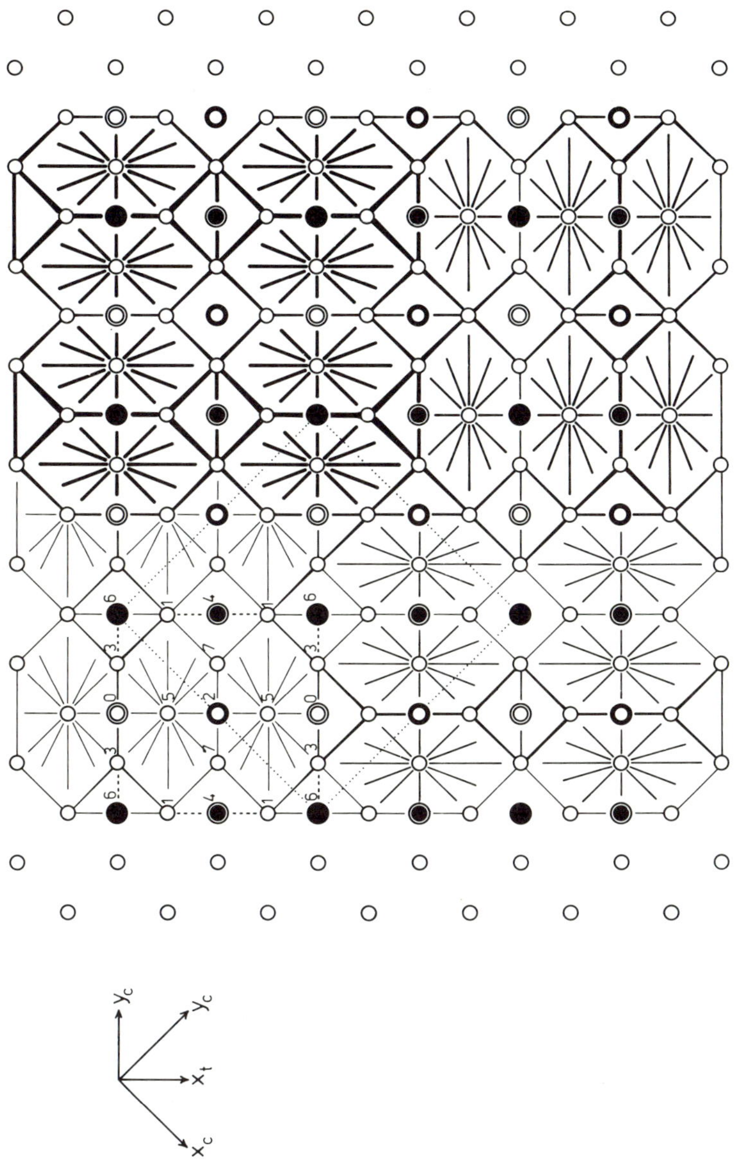

Fig. 26. γ-Fe_2SiO_4 (high-pressure spinel form) projected on (001) of the cubic cell; Fe and Si atoms only. *Large circle* = Si, *small circle* = Fe, heights in units of *c*/8. The f.c.c. unit cell is *dotted*, the equivalent b.c.t. unit cell ($\mathbf{c}_t \equiv \mathbf{c}_c$) outlined with *broken lines*. The $SiFe_{12}$ truncated tetrahedra are drawn: *top right*, Si at 3/4; *bottom right*, Si at 1/2; *bottom left*, Si at 1/4; *top left*, Si at 0. (Note that the Fe which *appear* to be at the centres of hexagonal faces are not. They form the lower edge of the truncated tetrahedron concerned.)

an O-centred B_3A tetrahedron. Figure 31 depicts the spinel γ-Fe_2SiO_4 in these terms (Fe_4 + Fe_3Si tetrahedra = stella quadrangula).

2.8.2 The Inverse Spinel Structure

An example of an inverse spinel with a well-determined structure is $TeLi_2O_4$[63)] (which is isostructural with $ZnLiNbO_4$). Being inverse, its formula may be written as $Li(TeLi)O_4$ to indicate that one half of the majority cations (Li) are now in the A sites and that the

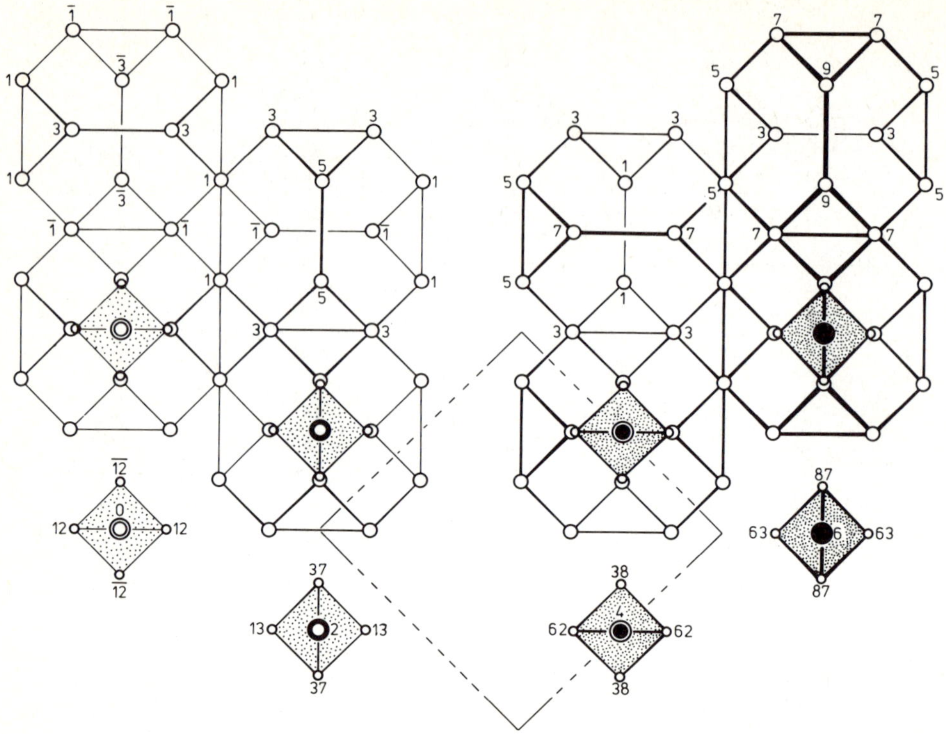

Fig. 27. γ-Fe_2SiO_4 projected on $(001)_{c,t}$. *Large, medium* and *small circles* are, respectively, Si, Fe and O. F.c.c. unit cell outlined. *Top,* Fe_{12} truncated tetrahedra; *bottom,* SiO_4 tetrahedra; *centre,* tetrahedra centering truncated tetrahedra. Single digits are atom heights in units of $c/8$; double digits are oxygen heights in units of $c/100$

other half plus the minority cations (Te) occupy the B sites of the normal spinel structure. The structure is shown in Fig. 32, which shows that the C9-like network of corner-connected tetrahedra (of atoms on B sites) is now composed of strictly alternating Li and Te atoms. The truncated tetrahedral interstices thus formed are centred by the remaining Li atoms (on the A sites).

The striking features of the structure (when compared with that of a normal spinel, Fig. 29) are (*a*) that the $Li(2)O_4$ tetrahedra are very tilted, so that the anion array is grossly distorted from the approximate cubic eutaxy observed in the normal spinel structure and (*b*) that, by contrast, the cation array is almost exactly that of the normal structure, and very regular indeed. [There is a small tetragonal distortion of the unit cell – symmetry $P4_122$ – which has $c/(\sqrt{2} \cdot a) = 0.9697$, compared with unity for the equivalent ratio of the cubic cell of the normal spinel structure.]

The anion-containing tetrahedra of cations (AB_3) are now of two types (instead of one, e.g. $MgAl_3$ in spinel itself), *viz.* Li_3Te and Li_2Te_2. These occur in equal numbers, two of each per $TeLi_2O_4$ formula unit. The oxygen atoms are therefore also of two, crystallographically distinct kinds (also in equal numbers): O(1) occupies the Li_2Te_2 tetrahedron [with l(Li–O) = 1.96_6 and 2.22_0 Å, and l(Te–O) = 1.93_6 and 2.04_2 Å]; O(2) is so far off centre in the Li_3Te tetrahedron that it is virtually 3-coordinate, almost in a

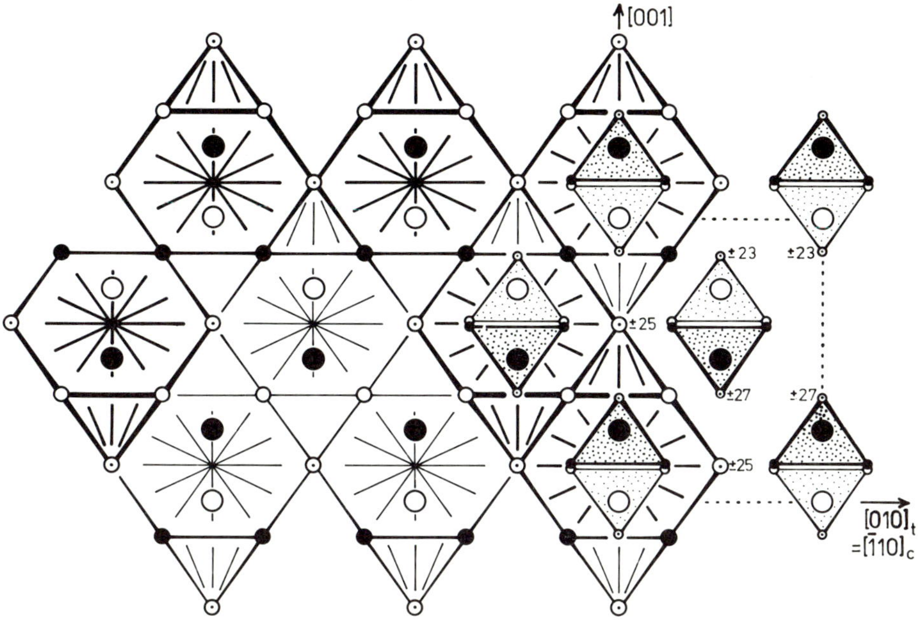

Fig. 28. γ-Fe_2SiO_4 projected on $(100)_t \equiv (100)_c$. *Circles* as in Fig. 27, but *open* at $x_t = 0$, *filled* at $x_t = 1/2$, *dotted* at $x_t \approx \pm 1/4$ (heights in units of $c/100$ in this case). B.c.t. unit cell outlined. *Left,* $SiFe_{12}$ truncated tetrahedra; *right,* SiO_4 tetrahedra; *centre,* both (the truncated tetrahedra being those about Si at 1/2)

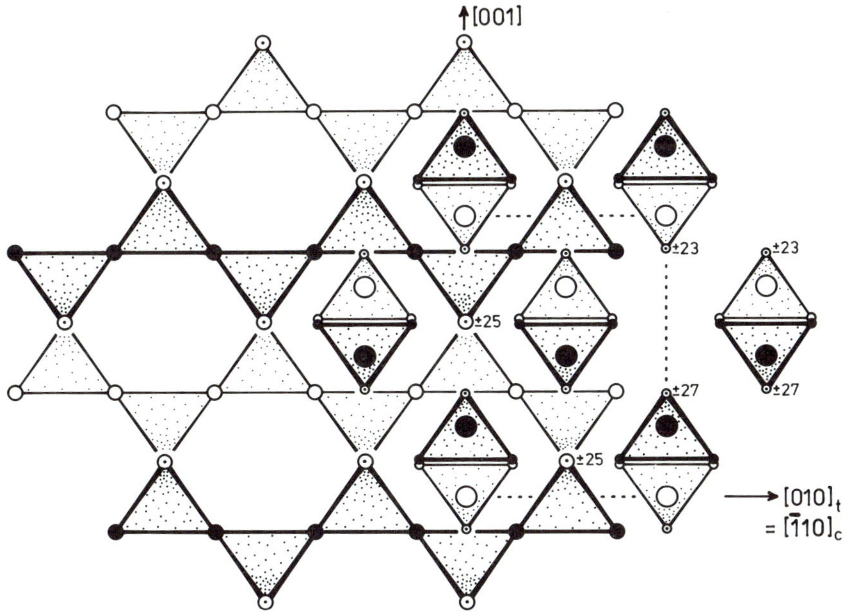

Fig. 29. γ-Fe_2SiO_4 projected on $(100)_t \equiv (110)_c$. This is perhaps the simplest description showing (*right* and *centre*) SiO_4 tetrahedra and (*left* and *centre*) the *C*9-like array of corner-connected Fe_4 tetrahedra that fills all the interstitial space between the $SiFe_{12}$ truncated tetrahedra

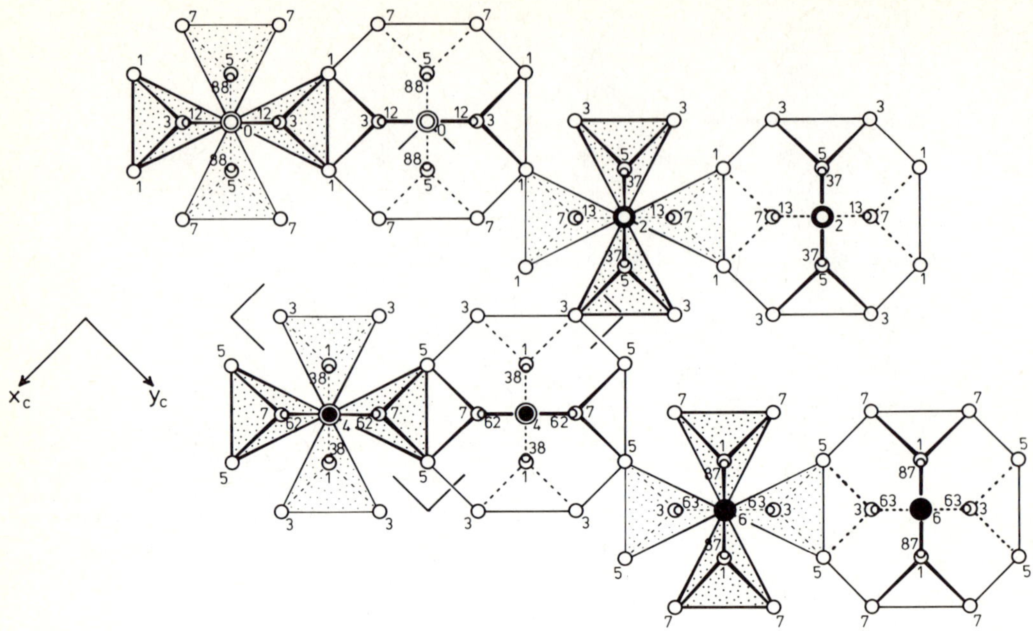

Fig. 30. γ-Fe_2SiO_4 projected on (001). *Large* and *medium circles* are, respectively, Si and Fe (heights in units of $c/8$); *small circles* are O (heights in $c/100$). Units at four different heights are drawn in pairs: in each case the one on the *right* is Si(+ O_4) centering an Fe_{12} truncated tetrahedron and the one on the *left* is the same unit drawn as OFe_3Si (*anion*-centred) tetrahedra in groups of four (within the truncated tetrahedron) and sharing a common Si atom. F.c.c. unit cell indicated

Li_2Te face of the tetrahedron, but with an additional, more distant Li atom [l(Te–O) = 1.91_0 Å, l(Li–O) = 1.86_1, 2.10_5 and 2.63_5 Å]. This last is related to the tilting of the Li(2)O_4 tetrahedron.

Apart from its orientation, the Li(2)O_4 tetrahedron is not very different from the AO_4 one in normal spinels. It is slightly distorted, edge lengths being d(O . . . O) = 2.78_1(2 ×), 2.78_5, 2.94_5(2 ×) and 3.16_5 Å; with two pairs of equal-length bonds, l[Li(2)–O(2)] = 1.86_1 and l[Li(2)–O(1)] = 1.96_6 Å. The coordination of the other cations by oxygen deviates considerably from that in normal spinels. The TeO_6 octahedron is rather badly deformed, with three pairs of equal-length bonds, l[Te–O(2)] = 1.91_0 and l[Te–O(1)] = 1.93_6 and 2.04_2 Å (a ratio of 1.000 : 1.014 : 1.069); i.e. TeO_{4+2}, moving towards tetrahedral coordination. The Li(1)O_6 octahedron is rather similarly deformed, but to a greater degree: the three pairs of (equal) bond lengths are now l[Li(1)–O(1)] = 2.22_0 and l[Li(1)–O(2)] = 2.10_5 and 2.63_5 Å (a ratio of 1.055 : 1.000 : 1.252). So, the coordination of Li(1) is really tetrahedral [as is that of Li(2)]. That is, at least in this inverted spinel, the normal structure ${}^{IV}A{}^{VI}B_2{}^{IV}X_4$ has changed to ${}^{IV}B({}^{VI}A{}^{IV}B){}^{III}X_2{}^{IV}X_2$ =

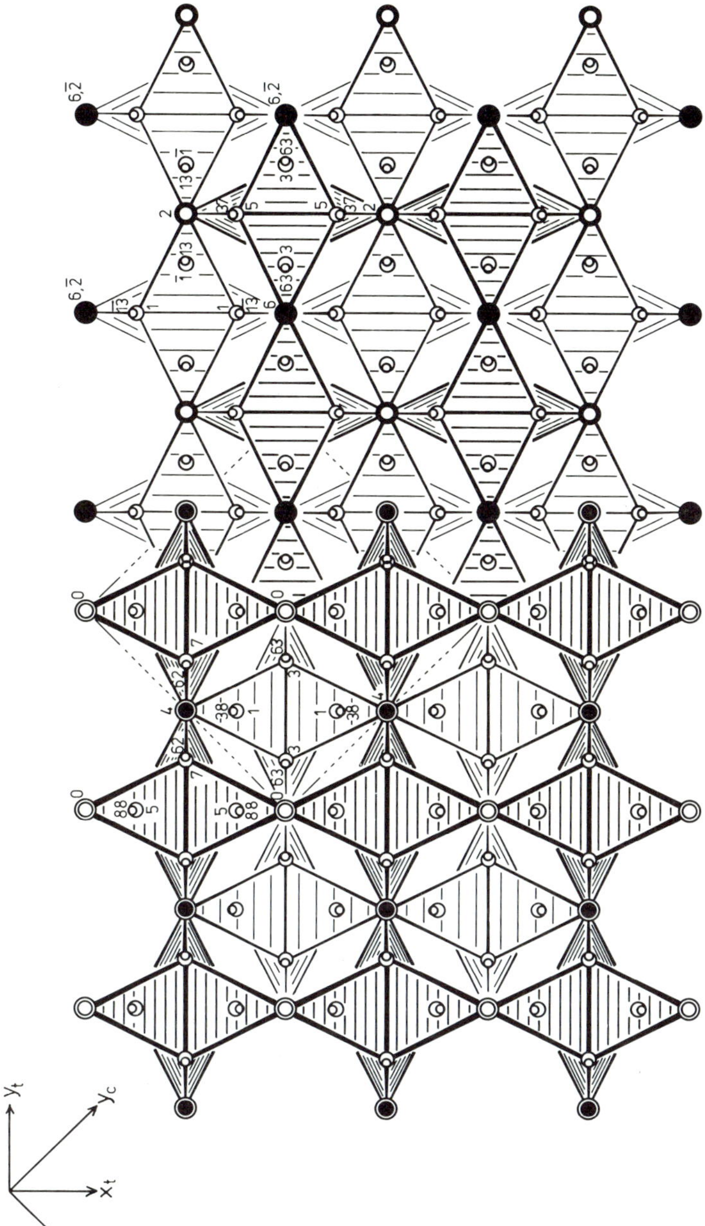

Fig. 31. γ-Fe_2SiO_4 projected on (001) and shown as stellae quadrangulae; $Fe_4(OSi)_4$ – an empty, central Fe tetrahedron capped on each face by an Si atom, with an O atom in each Fe_3Si tetrahedron. *Single digits* are atom heights in units of *c*/8, *double digits* are heights in units of *c*/100. To avoid confusion, in each half of the drawing (*left* and *right*) only one half of the stellae quadrangulae (and the relevant atoms) are drawn. An f.c.c. unit cell is outlined

$^{VI}A^{IV}B_2{}^{III}X_2{}^{IV}X_2$. The coordination numbers of the minority and majority cations have been interchanged. [Some of the interatomic distances given here differ from those originally reported[63] – it has occurred to us that one of the coordinates given in the original paper, and used by us, may be a misprint]

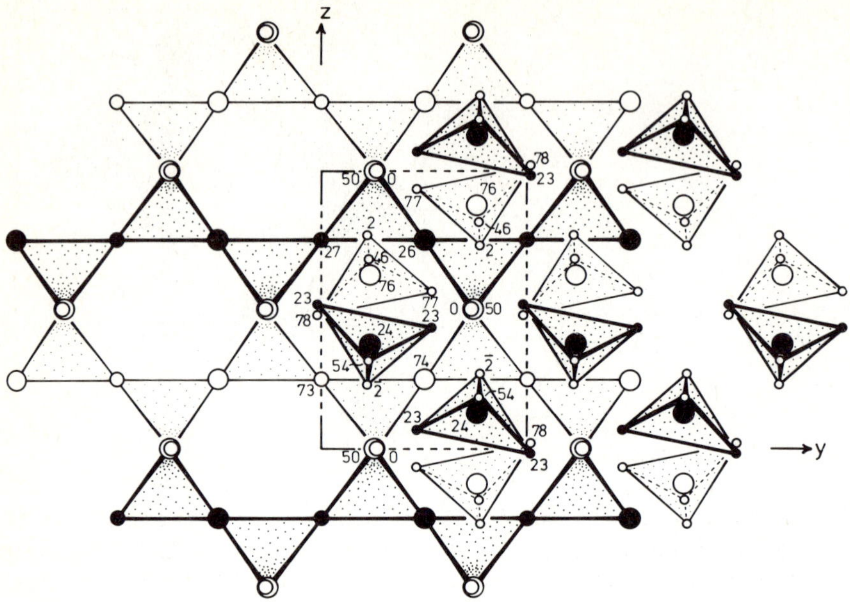

Fig. 32. The inverse spinel structure of $TeLi_2O_4$ projected on (100) of the tetragonal unit cell (compare Fig. 29). *Large circles* are Li and *medium circles* are Te (*filled* and *open* at heights of $\sim a/4$ and $\sim 3a/4$ respectively): *small circles* are O. Heights are in units of $a/100$

2.9 Structures Derived From the Cr_3Si (A 15) Structure; Geometrical Descriptions

2.9.1 The Garnet ($H3_1/S1_4$) Structure

This structure is very common, and its structural parameters have been the subject of considerable discussion[64–66]. We shall see later (Sect. 3.3) that (very much as in the example of the spinel structure, Sect. 3.2) the parameters of the garnet structure can be simply understood as a result of the oxygen atoms finding the appropriate positions within the cation array to produce the "correct" bond lengths.

Garnets, formula $A_3B_2C_3X_{12}$, are cubic; symmetry $Ia3d$, with A in 24(*c*) 1/8,0,1/4 etc., B in 16(*a*) 0,0,0 etc., C in 24(*d*) 3/8,0,1/4 etc. and X in 96(*h*) *x,y,z* etc. As Geller[67] has remarked, the cation array is a superstructure of the Cr_3Si(A 15) structure *viz.* $(A_{1.5}C_{1.5})B$. Just as for $MgCu_2$, the A 15 structure type is "topologically" close-packed. The space is divided into tetrahedra of the sorts A_3B, A_2BC, ABC_2 and A_2C_2, see Fig. 33. *All* the tetrahedra of the sort A_2BC are filled by X, so that all the X atoms have four coordination and the coordination numbers of A, B and C are eight, six and four respectively.

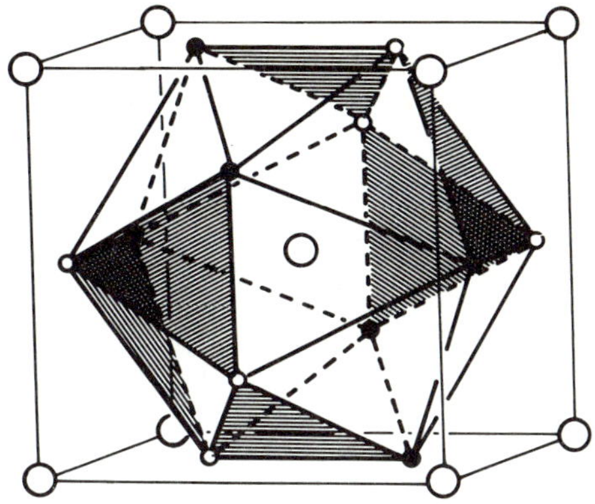

Fig. 33. The $(A_{1/2}C_{1/2})_3B$ array of cations in the garnet structure, $A_3B_2C_3X_{12}$, projected on {100}. Compare Fig. 34

2.9.2 *Ag_3PO_4 (H2₁) and $NaPt_3O_4$*

These structures are very closely related. The Ag_3P or Pt_3Na arrays are exactly as in the **Cr_3Si** structure [see Fig. 33 for $(Si_{1/2}Ca_{1/2})_3Al$ in garnet]. Thus in $NaPt_3O_4$ the Na atoms are at the corners and body centre of the cubic cell and surrounded by icosahedra of Pt. O atoms 1/4 and 3/4 of the way along each of the body diagonals are in the Pt_3 triangular faces shared by a pair of Pt_3Na tetrahedra and thus have trigonal bipyramidal coordination OPt_3Na_2[17]. The coordination of platinum is a PtO_4 square and of sodium a cube NaO_8.

The structure of Ag_3PO_4 is very closely related to that of Na_3PtO_4. The only difference is that oxygen atoms move alternately towards and away from the cell centre along the body diagonal (e.g. from 1/4,1/4,1/4 to *x,x,x*) so that now they are in PAg_3 tetrahedra, and instead of a PtO_4 square, one has a PO_4 tetrahedron. The symmetry is lowered from $Pm3n$ to $P\bar{4}3n$.

It is worth noting, in passing, that the $Pm3n$ structure has only one parameter (the cell edge, a), but two bond lengths Na–O and Pt–O that must be in the ratio $\sqrt{3/2}$. This is very similar to the situation in the ideal (cubic) perovskite structure (Sect. 2.2), and one might expect that the structure with this symmetry would be rare, the lower symmetry ($P\bar{4}3n$) structure being much more common; $Li_{1-y}Pt_3O_4$, for example has the $P\bar{4}3n$ structure[68]. This lower-symmetry structure has two parameters (a and x) that can be calculated in a straightforward way, just as for spinel (at least for the ideal composition, $y = 0$).

2.10 *Vesuvianite/Idocrase ($S2_3$)*

This complex structure has a large tetragonal unit cell with $a \approx 15.5$ Å, $c \approx 11.8$ Å and V ≈ 2.835 Å^3. There has been discussion about its exact composition, its correct space group (it seems not to be the same for all specimens), and about some aspects of the structure[69, 70]. For some specimens at least, this is now resolved[71].

17 Note that the occupied Pt_3 faces are eight of the 14 unshaded ones on the icosahedron of Fig. 33, those parallel to {111}. (The remaining 12 faces are parallel to {210}: they are the six shaded in Fig. 33, and six more equivalent ones that share ⟨100⟩ edges with those that are shaded)

From the beginning it was noted that the structure was dominated by large columnar elements of the garnet structure type[72]. The doubt has been about the remaining parts (between the garnet columns). The garnet structure itself is rather complicated; it has only recently been given a satisfactory description in terms of articulated, cation-centred polyhedra[66]. The vesuvianite structure is even more complex.

In Sect. 2.9.1 we described garnet as an oxygen-stuffed, A 15-like cation array with all the oxygens crystallographically equivalent. In vesuvianite[71] there are *eleven* crystallographically distinct oxygens [1 × 4(*c*), 1 × 8(*h*) and 9 × 16(*k*) of *P*4/*nnc;* a total of 156 O atoms]; but, qualitatively at least, a similarly simple description of the structure as an O-stuffed cation array can be given. Furthermore, it reveals a simple relationship between these two structure types. (Although, as already stated, it has long been known that the vesuvianite and garnet structures have much in common and are therefore related, the nature of the relationship has not, to our knowledge, previously been exposed.)

Figure 34 is a projection of the (A 15-like) cation array in garnet along a cube axis, after Andersson's description of A 15 itself[73]: *tetraederstern* share corners in the projection plane (and rhombic faces along the projection axis) to produce a framework within which rows of atoms are inserted (parallel to the projection axis). In A 15-type Cr_3Si these rows or rods are of Cr atoms, as are the central tetrahedra of the tetraederstern; the common vertices of the latter are Si atoms. In garnet, $A_3B_2C_3O_{12}$, the rods and the central tetrahedra are composed of alternating A and C atoms, and the common corners are B atoms. (For example, in $Ca_3Al_2Si_3O_{12}$ they are respectively alternating Ca and Si, and Al.)

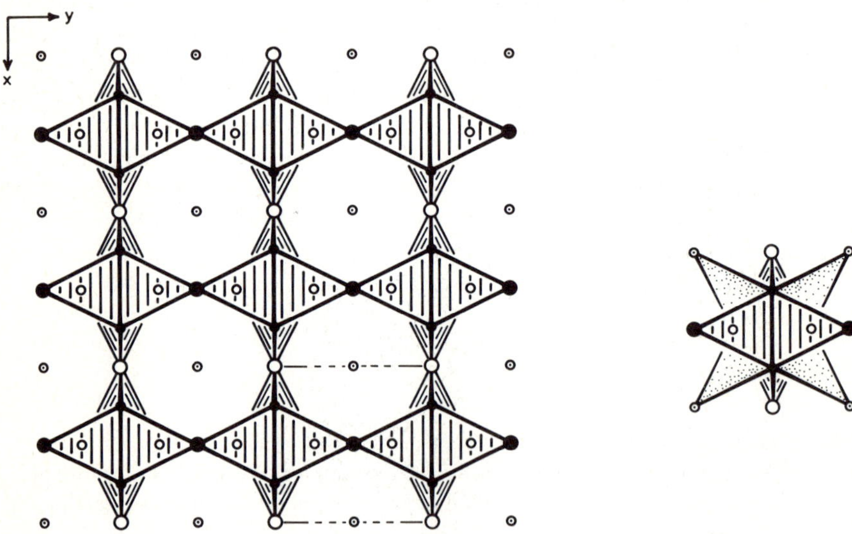

Fig. 34. The atom arrangement in Cr_3Si = A 15 structure type projected on {100}, after Andersson[73]. *Large* and *small circles* are Si and Cr respectively; *open* at $z/c = 0$, *filled* at $z/c = 1/2$ and, with *dotted centres* (Cr only), at $z/c = \pm 1/4$. On the *right* we show how these last atoms cap the "sloping" edges of the *central* tetrahedra (Cr_4) of the tetraederstern. One unit cell indicated

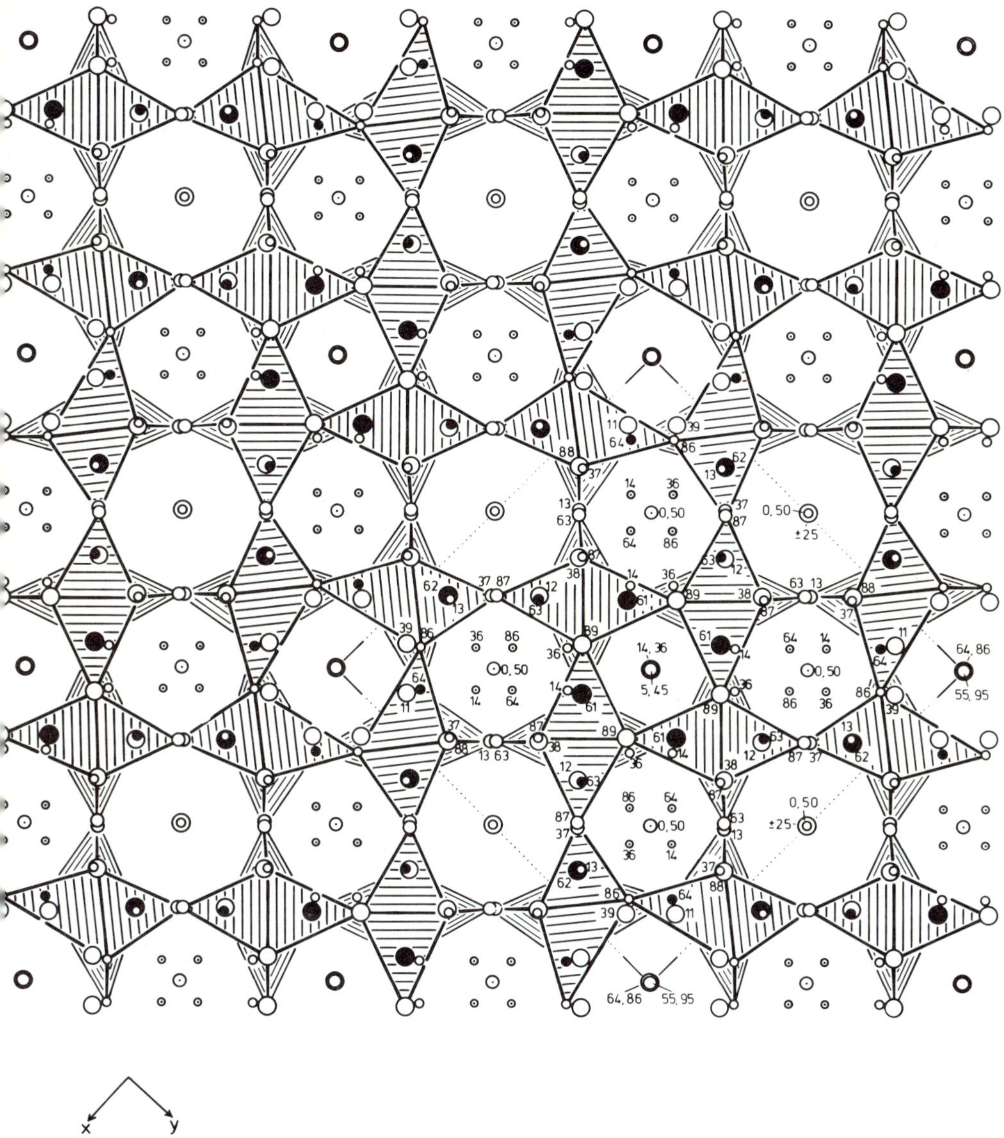

Fig. 35. The cation array in vesuvianite (idocrase) projected on (001). In the framework the *large circles* are Ca, *small circles* are Si and *medium circles* are (Al, Fe). In the tunnels the *medium open circles* are Fe (with Ca in square tunnels), those with *dotted centres* are Al (*medium*) and OH (*small*) (in the pentagonal tunnels). The hexagonal tunnels are occupied by CaSi rods. Only one set of tetraederstern ($5/8 \times \mathbf{c} \lesssim z \lesssim 7/8 \times \mathbf{c}$) is drawn, and only the relevant appropriate atoms (at $z/c \approx 5/8$) represented by *filled circles.* This is to assist comparison with Fig. 36, the equivalent projection of Mo_3CoSi. (But note that the **c** axis of vesuvianite is equivalent to $2 \times \mathbf{c}$ of Mo_3CoSi.) Atom heights are in units of $c/100$. One unit cell is outlined, but its origin shifted to correspond to that in Fig. 36

Figure 35 shows the cation array in vesuvianite[71] similarly projected along the tetragonal axis. (Note that this axis has the same length in both structures.) Again a framework of tetraederstern is readily recognised (although they are slightly distorted), but now some of the vertex-sharing (in A 15) has been replaced by edge-sharing. Each column of four *corner*-shared tetraederstern is the same as in one unit cell (in cross-section) of garnet. But the columns in vesuvianite are joined by edge-sharing of tetraederstern whereas, in garnet, they are corner-sharing. The change (garnet → vesuvianite) is a double crystallographic shear (*CS*) operation: there are two, mutually orthogonal sets of (parallel, regularly-spaced) *CS* planes in the latter. Hence, as far as the cation framework is concerned, vesuvianite is derived from garnet in the same way as the well known column (= "block") structures of various niobium oxides etc. are derived from the $\mathbf{ReO_3}$ type[52]. Andersson[73] has described exactly this relationship in the (oxygen-free) alloy structures. The vesuvianite framework appears in the structure of Mo_3CoSi, Fig. 36. In the former the central tetrahedra (in the tetraederstern) are again (cf. garnet) alternating Ca and Si; the shared vertices are (Al, Fe). (In $Ca_3Al_2Si_3O_{12}$ they are just Al.) The stoichiometry of the framework has changed from e.g. $Ca_{32}Al_{32}Si_{32}$ (garnet) to $Ca_{32}(Al, Fe)_{16}Si_{32}$ (vesuvianite), i.e. $A_{32}B_{32}C_{32}$ to $A_{32}B_{16}C_{32}$. Only the shared corners (B atoms) are reduced in number by the *CS* operations.

We come now to the [001] rods of cations in the interstices of the framework. In A 15 Cr_3Si these are ....CrCrCrCr... rows; in garnet they are ...ACAC... (e.g. ...Ca Si Ca Si...); in vesuvianite the analogy is not always exact. The structure contains three different sorts of tunnel: per unit cell, two are hexagonal (as in A 15 and garnet), four are pentagonal, and two are square – see Fig. 35. They are filled in the following ways.

(i) Those at $\pm(3/4,1/4,z)$ are exactly as in garnet (hexagonal), and so is the rod: alternating Ca and Si atoms are at $\pm 1/4$ and 0,1/2 respectively; i.e. at intervals of $\mathbf{c}/4$.

(ii) Those at $0,0,z$; $0,1/2,z$; $1/2,0,z$ and $1/2,1/2,z$ are collapsed by *one* of the *CS* operations, so that they are pentagonal in cross-section. The rods within them consist of two Al atoms at $z/c = 0,1/2$: but also the hydroxyl groups at $z/c = \pm(1/4 \pm 0.11)$ will yield two H_2^{2+} pairs (which are effective cations[74]) at $\pm 1/4$. So, in this case too, there are cations at intervals of $c/4$ along the rods. (But they are alternating H_2 and Al instead of Ca and Si.)

(iii) At $\pm(1/4,1/4,z)$ *two CS* planes intersect, and the tunnels are even more collapsed (compared with garnet), now being square in cross-section. On $(1/4,1/4,z)$ there are Ca sites at $z/c = 3/4 \pm 0.09$ and Fe sites at $z/c = 0.55, 0.95$; but all have a 50% occupancy. For the similar rods at $-(1/4,1/4,z)$ these heights have to be multiplied by -1. Thus, while there are still four cation sites in the repeat length of these rods (as in garnet) they are (*a*) only half-occupied and (*b*) not equidistant. (Also, and in addition, there are two oxygen atoms in each of these rods.)

In spite of these last, minor differences the relationship to garnet is close enough to warrant interest and description. The cation rods in the tunnels change from $4 \times Ca_2Si_2$ (all hexagonal) in (Ca, Al, Si) garnet to $2 \times Ca_2Si_2$ (hexagonal), $4 \times (H_2)_2Al_2$ (pentagonal) and $2 \times (Ca_{1/2})_2(Fe_{1/2})_2[O_2]$ (square) in vesuvianite. Ignoring the collapse due to *CS*, the axial relationships are $\mathbf{a}_v = \mathbf{a}_g - \mathbf{b}_g$, $\mathbf{b}_v = \mathbf{a}_g + \mathbf{b}_g$, $\mathbf{c}_v = \mathbf{c}_g$; i.e. $\sim 1\bar{1}0/110/001$: compare $a_g = b_g = c_g \approx 11.8$ Å with $a_v = b_v \approx 15.5_0 = \sqrt{2} \times 10.9_6$ Å, $c_v \approx 11.79$ Å; $(a_v/\sqrt{2})/c_v = 0.93$, i.e. the *CS* collapse vector is $\sim 0.07\, a_v\langle 110\rangle$ whereas the ideal value is $1/8 = 0.125\, a_v\langle 110\rangle$. [That it is less than the ideal value – which assumes no expansion

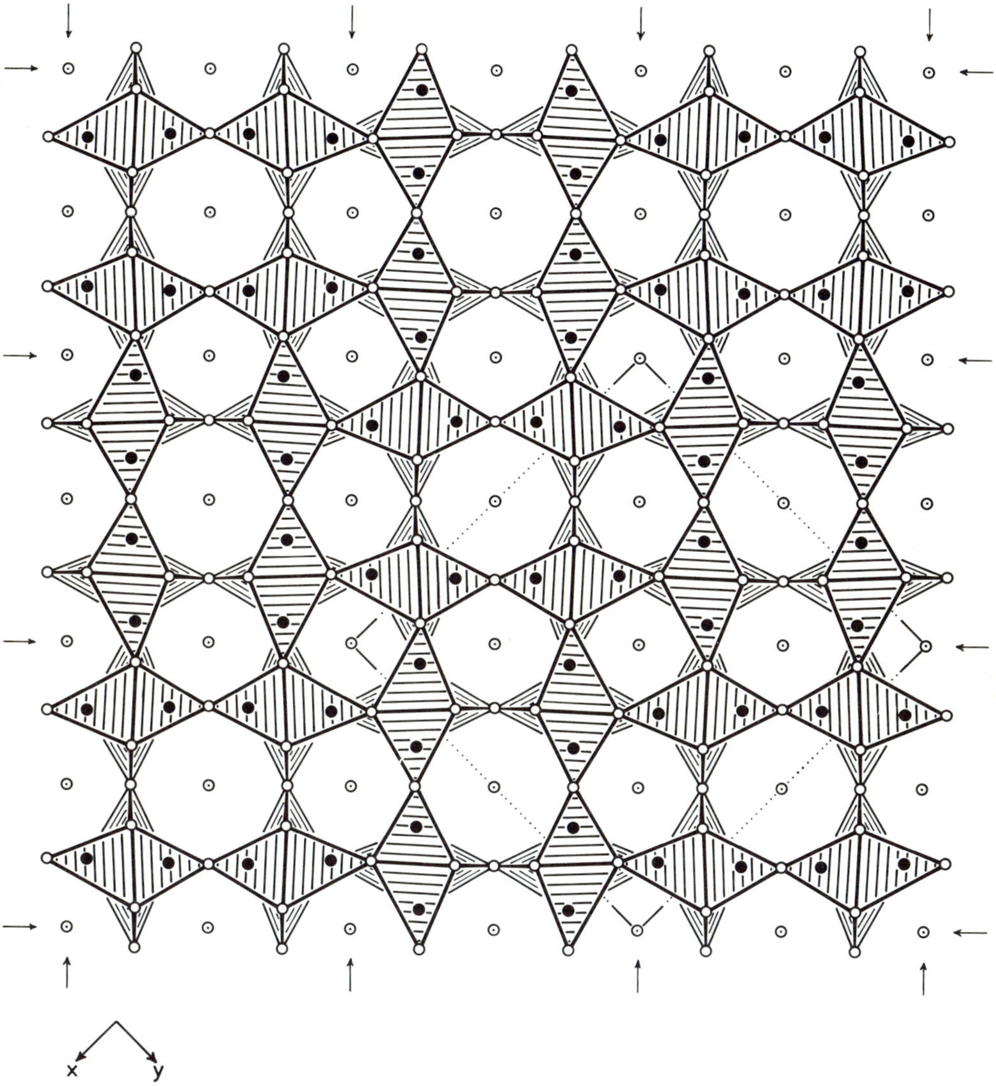

Fig. 36. The structure of Mo_3CoSi projected on (001), after Andersson[73]. Compare Fig. 35. There is mixed occupancy of most sites; and this is not shown. *Open circles* are atoms at $z/c \approx 1/2$ (1/2 or ± 0.49); *filled circles* are atoms at $z/c \approx 0$ (0 or ± 0.01); *dotted circles* are at $z/c = \pm 1/4$. Compare Fig. 34 and Fig. 35. One unit cell is outlined. *Light arrows* at the periphery of the drawing indicate the *CS* planes that transform $\mathbf{Cr_3Si}$ (Fig. 34) to $\mathbf{Mo_3CoSi}$. Note the "block"/column structure that results; each column being 2×2 tetraederstern in cross-section (i.e. 2×2 unit cells of A 15 $\equiv 1 \times 1$ of garnet). The columns are joined (at the *CS* planes) by edge-sharing of adjacent tetraederstern; cf. corner-sharing in A 15

due to repulsions causing strain and/or change in shape of coordination polyhedra – is the normal state of affairs in *CS*.]

2.11 Structures Based on the Mn_5Si_3 ($D8_8$) Alloy Type

2.11.1 Hydroxyapatite, $Ca_5(PO_4)_3 \cdot OH$ ($H5_7$)

This is a rather complex structure and versatile, e.g. OH^- can be replaced by F^-, Cl^-, 1/2 CO_3^{2-}. It cannot be adequately described in terms of articulated coordination polyhedra, unlike the structures dealt with in e.g. Sects. 2.7, 2.8. Like those in some earlier sections (e.g. Sects. 2.3–2.5) the coordination polyhedra of the small cations are obvious, PO_4 tetrahedra, but not the anion polyhedra about the larger Ca ions. Wells[4)] does not mention the structure; Wyckoff does not describe it except by a diagram.

The structure (Holly Springs hydroxyapatite)[75)] is hexagonal, and is drawn in Fig. 37. There are two crystallographically distinct Ca sites: in space group $P6_3/m$

Ca(1) in 4(f) at $\pm(1/3,2/3,z)$,

and Ca(2) in 6(h) around the origin of the unit cell.

The (frequently recognised) anion coordination of Ca(1) (see lower part of Fig. 37) is CaO_6: between an octahedron and a trigonal prism (a "metaprism"). The anion coordination of Ca(2) is irregular and ill-defined.

However, the structure is perfectly straightforward when regarded as an anion-stuffed alloy. The alloy part (i.e. the cation array) is shown in the upper part of Fig. 37: Ca_5P_3, virtually identical to the $\mathbf{Mn_5Si_3}$ ($D8_8$) structure type[18]. The $\mathbf{Mn_5Si_3}$ structure type has itself been rather difficult to comprehend and describe: we follow the recent, novel description by Nyman and Andersson[76)].

The Ca(2) atoms, arrayed around 00z, form columns of face-sharing Ca_6 octahedra along the **c**-axis, as in h.c.p. (see especially the top of Fig. 37). All the *edges* of these octahedra are capped by additional atoms ("triangular capping"): the sloping edges by the Ca(1) atoms, forming almost equilateral triangles (edge lengths 4.03, 4.17, 3.95 Å); those edges parallel to (0001) and common to two octahedra by (smaller) Patoms, forming almost isosceles triangles [edge lengths 4.08 (Ca...Ca), 3.08 (Ca...P), 3.29 Å (Ca...P)]. The columns of $Ca(2)_6$ octahedra are linked via the capping Ca(1) atoms, each of the latter being common to three columns.

All the oxygens forming the PO_4 tetrahedra are in PCa_3 tetrahedra of two types, $OPCa(1)Ca(2)_2$ and $OPCa(1)_2Ca(2)$; OH groups occupy *half* the 4(e) sites along [00z],

18 Since this was written we have discovered that "a resemblance between the two structures" is mentioned (although not described) by Schubert[3)]. Furthermore, it has now been brought to our attention that the same relationship between the apatite and Mn_5Si_3 structure types was clearly spelled out by Wondratschek, Merker and Schubert[106)] twenty years ago. At the same time the relation between the structures of $Bi_4(SiO_4)_3$ and Th_3P_4 was also pointed out. And subsequently[107)] two of these authors determined the structure of $Pb_8(PO_4)_5I = 2Pb_4[(PO_4)_{2\,1/2}I_{1/2}]_3$, which is similar to that of eulytine. (The eulytine/Th_3P_4 relation is also included in Table 3.) We are happy to acknowledge the priority of these descriptions, and are grateful to Professor Wondratschek for providing us with the details

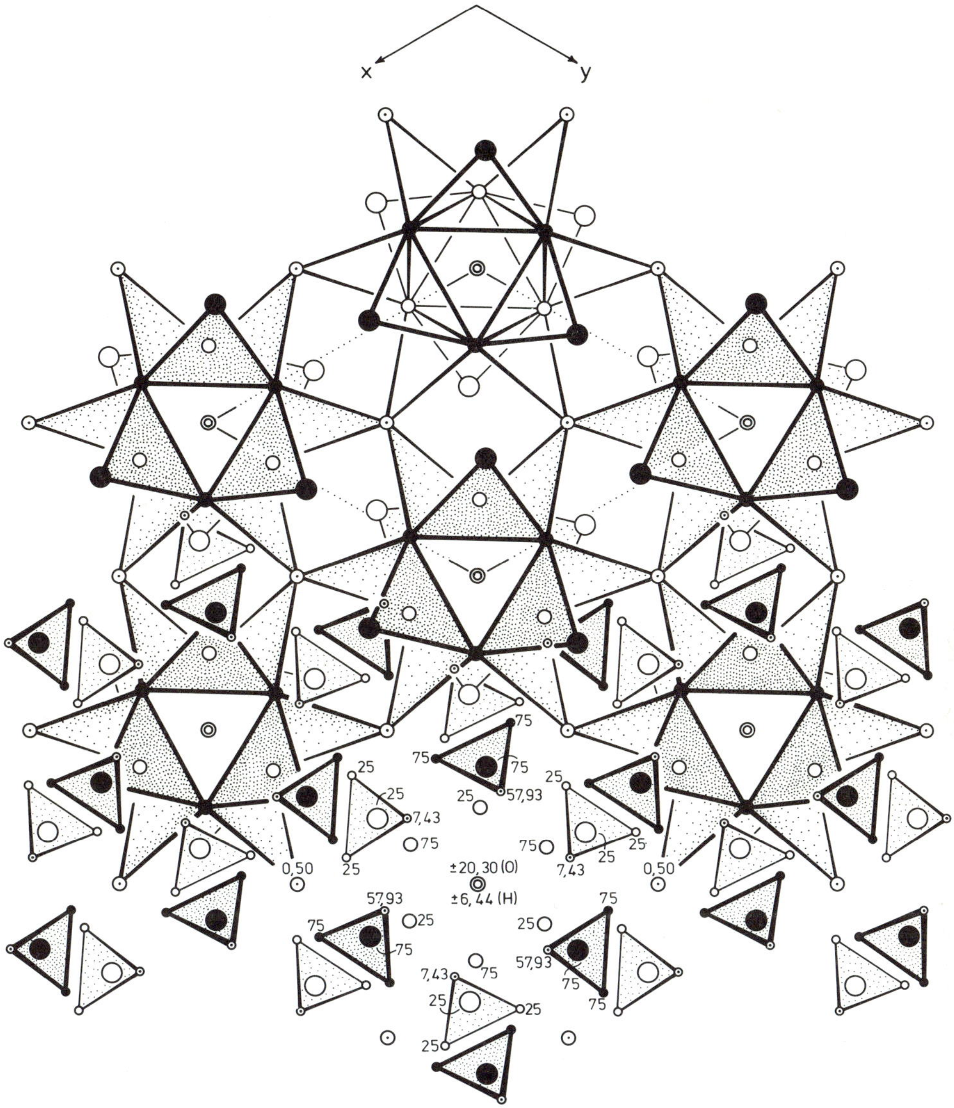

Fig. 37. The structure of hydroxyapatite, $Ca_5(PO_4)_3OH$, projected on (0001). *Large, medium* and *small circles* are respectively P, Ca and O. The *double circles* are OH groups lying along [000*z*]. Atom heights are in units of *c*/100: *open* (except OH) at $z = 1/4$, *filled* at $z = 3/4$, *dotted* at $z \approx 0,1/2$. The unit cell is outlined. *Top,* cations only, emphasising the Nyman-Andersson description of $\mathbf{Mn_5Si_3}$($D8_8$) as columns of face-sharing $\mathbf{Mn_6}$ octahedra (along [000*z*]) capped on sloping edges by **Mn** (forming $\mathbf{Mn_3}$ triangles) and on "horizontal" edges [//(0001)] by **Si** atoms (forming $\mathbf{Mn_2Si}$ triangles) – in the present case Ca_6, Ca_3 and Ca_2P respectively. (In apatite these "stars" at [000*z*] are slightly rotated – all clockwise – compared with those in $\mathbf{Mn_5Si_3}$, by about 4°.) *Bottom,* PO_4 tetrahedra and Ca atoms. *Centre,* Ca_3 and Ca_2P triangles plus PO_4 tetrahedra. Note that only half the O, H sites at 000*z* are occupied: these oxygens lie in Ca_3H tetrahedra. (In fluorapatite OH is replaced by F, in octahedral faces i.e. Ca_3 triangles: in chlorapatite by Cl at 0,0,0.444, i.e. off-centre in $ClCa_{3+3}$ octahedra. CO_3 groups may also (just!) fit into these tunnels so that each O is in a Ca_3C tetrahedron

putting these oxygen atoms into HCa_3 tetrahedra. {In fluorapatites they are replaced by F ions along [00z] – now centred in Ca_3 triangles; in chlorapatite the Cl ion is at (00z), just off centre in the Ca_6 octahedra (coordination $ClCa_{3+3}$). Cf. ref.[77]. In carbonate apatites the CO_3 groups are also centred at (00z), and oriented so that the oxygens are in $OCCa_3$ tetrahedra.}

Thus, although this structure cannot be described or understood in terms of any regular anion array or any connected array of cation-centred polyhedra, it is readily and logically visualised as an alloy framework with anions inserted into appropriate (usually tetrahedral) interstices. It must be pointed out, however, that there is a very small distortion of the $\mathbf{Mn_5Si_3}$ structure to give the cation array in apatite. This consists of a slight rotation of each of the columns ("rosettes") at [00z] – in Fig. 37 clockwise by about 4°. This affects the symmetry, but hardly the geometry.

One interesting isostructural compound is $Y_5(SiO_4)_3N$: its Y_5Si_3 array is the same as that in the alloy Y_5Si_3 itself ($D\,8_8$)!

2.11.2 Barium Per-Rhenates, $Ba_5(ReO_5)_3X$, With $X = O_2$, F or $(CO_3)_{1/2}$

These are referred to as "apatite-like structures"[78–80] but instead of PO_4 tetrahedra they contain ReO_5 square pyramids. Their "apatite-like" feature is the Ba_5Re_3 array (plus the X group at 00z). The space group is, in each case, $P6_3cm$ (cf. $P6_3/m$ for apatite): Ba atoms occupy the **Mn** positions of $\mathbf{Mn_5Si_3}$ (Ca in apatite) and Re atoms the **Si** positions (P in apatite). The X atom or group is (as in hydroxy-, fluor- and chlor-apatite) in the tunnels/columns of face-sharing $Ba_6(Ca_6)$ octahedra at [00z]. Four of the five oxygen atoms of the ReO_5 groups are again in tetrahedral coordination (as they are in apatite), $OReBa_3$; the fifth (the apical oxygen of the O_5 pyramid) is only 2-coordinate, OReBa. The slight rotation of each of the columns ("rosettes") at [00z], observed in apatite, is absent in these structures. That is, the Ba_5Re_3 array is closer than the Ca_5P_3 array to the atom array in Mn_5Si_3.

Thus, the difference between these structures and the apatites is simply in the choice of tetrahedral interstices in the alloy which are used to accommodate the anions. This determines the type of cation-centred polyhedron for the minority cation. (We ignore the trivial variations in anions inserted into the [00z] tunnels.) This suggests that the substitution of CO_3^{2-} for PO_4^{3-} in apatites may well be a less radical change than is usually assumed (when anion packing is accepted as being dominant).

2.11.3 Nasonite, $Pb_6Ca_4(Si_2O_7)_3Cl_2$, and $Pb_5Ge_3O_{11}$

These structures are very simply related to that of apatite, as previously pointed out by Newnham et al.[81]. They are worthy of a mention here because, once again, the relation is more clearly seen if, in the first instance, the cation arrays are compared and the anions ignored.

The relations involve introducing into the apatite structure mirror planes parallel to (0001) at $z/c = |0|_{\text{mod }1}$ or $|1/2|_{\text{mod }1}$. The ("Mn_6") octahedra that were at 00z, and centred at the mirror plane level, are thus converted to trigonal prisms. In nasonite these mirrors are at intervals of $z/c = 1$ (of the apatite structure), so that octahedra and trigonal prisms (of Pb) alternate along **c.** In $Pb_5Ge_3O_{11}$ the repeat unit is 2 octahedra + 1 trigonal prism. All have $a \approx 10$ Å, but the **c** axes are (for apatite, nasonite and $Pb_5Ge_3O_{11}$ respectively) ~7 Å, 13 Å and 10 Å; ratios of 1:2:1 1/2.

The mirror operation [which may also be regarded as a rotation of (0001) layers of the structure by 60° about the 6_3 axis] puts 2 oxygens – those at ±7 or ±43 in Fig. 37 – very close together (separation = $\Delta z = 0.14c \approx 1$ Å). They are replaced by a single oxygen at an intermediate position; and a pair of PO_4 tetrahedra (in apatite) thus become corner-connected, forming Si_2O_7 groups in nasonite, Ge_2O_7 groups in $Pb_5Ge_3O_{11}$. The stoichiometries (ignoring the Cl/tunnel site in chlor-apatite and nasonite) are then $A_{10}[(BO_4)_2]_3$, $A_{10}[B_2O_7]_3$ and $A_{10}[(BO_4)_2][B_2O_7]_2$ respectively; i.e. $A_{10}B_6O_{24}$, $A_{10}B_6O_{21}$ and $A_{10}B_6O_{22}$. The basal-plane-twinning operation is presumably to accommodate the varying anion/cation stoichiometric ratio, O/B.

Additional, albeit metastable, lead germanates have recently been reported; their structures are based on the same principle, but with different spacings between the (0001) mirror planes[82)].

2.12 Tetrahedral Framework Structures

The structures discussed above – those in which the cation arrays are alloy structures – are those in which also the oxygen atoms are in 3 or 4 (or occasionally 5 or 6) coordination. In oxides such as SiO_2, in which oxygen has 2-coordination (so that there are no anion-centred coordination polyhedra) and silicon 4-coordination, each Si atom necessarily has but four nearest neighbour Si atoms, so that the Si arrangement is that of a four-connected net. A simple example is that of cristobalite in which the silicon arrangement is that of diamond (or Si itself). In other (tetrahedral) framework structures, such as the alkali and alkaline earth alumino-silicates, the $(Al, Si)O_2$ framework similarly corresponds to an (Al, Si) four-connected net with the most electropositive atoms in interstices of this framework. A simple example is $KAlO_2$ (stuffed cristobalite structure) in which the KAl arrangement is that of the Zintl phase (B 32) structure of e.g. LiAl, which can be described as Li inserted into a diamond-like framework of Al.

In discussing the topology of such structures, it is particularly appropriate then to focus on the structure of the four-connected nets of metal atoms – a procedure in tune with the philosophy of this paper. It is in fact an approach that has already been actively developed elsewhere (see e.g. Ref.[83)]), so we do not pursue this topic further here, other than to include in Table 3 some examples of alumino-silicates in which the metal atom arrangement is topologically equivalent to corresponding network alloy types.

3 Interatomic Distances and Structural Parameters

In many of the structures we have described, the metal atoms are in high symmetry positions, often at the positions of invariant lattice complexes. In the latter cases the problem of completely specifying the structure then becomes one of locating the anions in their coordination polyhedra (of cations). Regarded from this viewpoint, the structural parameters of some of the simpler structures become immediately understandable as just those required to provide appropriate (cation-anion) bond lengths in the crystal. We emphasise this elementary point, and provide a few examples below, as an alternative to the more common, but confusing, discussions in terms of distortions of the anion arrangements in cation-centred polyhedra. It is felt, too, that our discussion will provide a useful demonstration of the point that some apparently complex crystal structures, such as that of garnet, are of just the above sort; and that elaborate calculations using many-parameter interatomic potentials do not, despite frequent claims to the contrary, provide any proof of the appropriateness of those potentials.

For our present purposes, simple crystal structures might be segregated into three groups. The first of these is comprised of those few structures with even fewer parameters than independent bond lengths; examples already encountered above include the (cubic) perovskite, pyrochlore and Na_3PtO_4 structures.

The second group consists of those structures that have the same number of parameters as independent bond lengths. In this case the structural parameters are determined by requiring the bond lengths to have their "normal" values. For oxides this can be done by using appropriate tables of bond lengths, such as those of Shannon and Prewitt[84]. Conspicuous and important examples of this second class are the spinel and garnet structures.

The third and by far the largest class is that of structures with more parameters than bond lengths. A calculation of the structure must now include a consideration of non-bonded distances, and in favourable cases might be expected to provide insight into the relative importance of the different kinds of non-bonded interactions in the crystal (mainly repulsions). Even here, caution must be exercised. In a number of cases it has been shown that non-bonded repulsions can be successfully simulated by the simple device of maximising the crystal volume subject to the constraint of fixed bond lengths[9]. Only when this ploy fails will it be necessary to enquire more closely into the nature of interatomic forces.

3.1 The Pyrochlore Structure

In Sect. 2.1.1.3 we considered the geometry of the pyrochlore, $A_2B_2O(1)_6O(2)$, structure. We now analyse it quantitatively.

The O(1) atom at x,1/8,1/8 has as nearest neighbours two equidistant A atoms (at 1/4,0,1/4 and 1/4,1/4,0) and two equidistant B atoms (at 1/2,1/4,1/4 and 1/2,0,0) (cf. Sect. 2.1.1.3). If now α and β represent the A–O and B–O bond lengths, we have

$$\alpha^2 = a^2[(1/4 - x)^2 + 1/32] \tag{1}$$

$$\beta^2 = a^2[(1/2 - x)^2 + 1/32] \tag{2}$$

which may be solved to give a and x in terms of α and β:

$$a^2 = (16/3)[\alpha^2 + \beta^2 + (8\alpha^2\beta^2 - 2\alpha^4 - 2\beta^4)^{1/2}] \tag{3}$$

$$x = 3/8 + 2(\alpha^2 - \beta^2)/a^2 \tag{4}$$

Equation (4) explains immediately the empirical observations made (see e.g. Barker et al.[85]) of the dependence on relative atomic sizes of the parameter x.

However care must be taken in using bond lengths from tables of ionic radii to solve for a and x directly from Eqs. (3) and (4). This is because, in addition to the A–O(1) bond there is also an A–O(2) bond. In fact the coordinations of the various atoms are $BO(1)_6$, $AO(1)_6O(2)_2$, $O(1)A_2B_2$ and $O(2)A_4$. The A–O bond lengths are now best correlated to bond strengths using the bond valence sum rule which states that the sum of the bond strengths at each atom is equal to the atomic valence[86]. We start with the $BO(1)_6$ polyhedron for which the bond strength is $s = Z_B/6$ (here Z_B is the formal valence of B, e.g. 4 for Ti). Likewise the strengths of the bonds in the $O(2)A_4$ tetrahedron are $s = 1/2$. Consider now the $O(1)A_2B_2$ tetrahedron: we have for the strength of the O(1)–A bond, $s = 1 - Z_B/6$ [this will make the sum of the bond strengths at O(1) equal to 2]. As a check we consider the $AO(1)_6O(2)_2$ polyhedron. The sum of the strengths of six A–O(1) bonds and two A–O(2) bonds is $6(1 - Z_B/6) + (2 \cdot 1/2) = 7 - Z_B = Z_A$ (the valence of A).

Thus for α we must take the bond length appropriate for a bond strength $s = 1 - Z_B/6$. To illustrate, we take at random an example of a pyrochlore with a well-determined structure[87], $Y_2Ti_2O_7$ with $Z_B = 4$, so that the strength of the bond of length α is 1/3. This would correspond to the bond length found for yttrium forming nine equal bonds to oxygen; it can be found from tables of ionic radii[84] or from bond strength-bond length correlations[86]. (It is worth emphasising that bond lengths determined from sums of ionic radii for given coordination numbers are just bond lengths for certain rational bond strengths. When there are coordination figures with more than one kind of bond, in particular mixed-anion compounds such as oxy-nitrides, bond strength considerations of the kind given above are necessary for correct results[25].)

From tables of ionic radii[84] we find for $Y_2Ti_2O_7$ $\alpha = 2.455$ Å, $\beta = 1.958$ Å and calculate from Eqs. (3) and (4) (observed values in parentheses):

$$a = 10.082\ (10.090)\ \text{Å}\ ,$$

$$x = 0.419\ (0.420)\ .$$

Similar calculations for other compounds yield comparable results.

It must be pointed out, however, that there are three different metal-oxygen bonds of fixed bond strength in the structure. As well as α and β there is the A–O(2) bond of length α′ with

$$\alpha' = \sqrt{3}\,a/8\ .$$

As a is already determined by α and β, α′ is not necessarily the optimum value for that particular bond. This is analogous to the situation in Na_3PtO_4 (Sect. 2.9.2) and in the ideal perovskite structure (Sect. 2.2). To relieve the overdetermination, the symmetry must be lowered; and it is perhaps significant that lower symmetries have been reported for some pyrochlores (e.g. $Cd_2Nb_2O_7$)[104].

3.2 The Spinel $H1_1$ Structure

As discussed above in Sect. 2.8, in the spinel (AB_2X_4) structure each anion (X) is coordinated by one A and three equidistant B atoms. The B...B distances are $a/\sqrt{8}$ and A...B are $a\sqrt{11}/8$. An X atom at u,u,u ($u \approx 3/8$) has neighbours A at 1/4,1/4,1/4 and three B at 1/8,3/8,3/8; 3/8,1/8,3/8 and 3/8,3/8,1/8. It is elementary to derive the following expressions[19] relating the structural parameters a und u to the bond lengths $\alpha = l(A\text{–}X)$ and $\beta = l(B\text{–}X)$ and to their ratio $\varrho = \beta/\alpha$:

$$u = 3/8 + [(33\,\varrho^2 - 8)^{1/2} - (3\,\varrho^2 + 2)]/[24(\varrho^2 - 1)] \tag{5}$$

$$a = \alpha/[\sqrt{3}\,(u - 1/4)] \tag{6}$$

Now, it is very well established that bond lengths are remarkably constant from compound to compound for a given coordination number or better, in more complex structures, for a given bond strength. Thus, from well-established tables of bond lengths for oxides[84)] we can obtain α and β appropriate to any structure. Equation (5) tells us that u depends only on the ratio of the bond lengths, so that u (and hence the details of the oxygen packing) represent nothing more than the response of the structure to the necessity of providing to correct bond lengths.

For example using the ionic radii of Shannon and Prewitt[84)] we calculate (observed values are in parentheses):
for

$MgAl_2O_4$: $a = 8.09\ (8.08)$ Å, $u = 0.389\ (0.387)$,

and for

γ-Fe_2SiO_4: $a = 8.28\ (8.23)$ Å, $u = 0.365\ (0.366)$.

We have already mentioned (Sect. 2.8) that perfect cubic eutaxy of the anions in the spinel structure requires $u = 3/8$. Equation (5) shows that this occurs when $\varrho = \sqrt{4/3} = 1.155$. Another interesting case is that for "tetrahedral bond angles", i.e. $B\hat{X}B = A\hat{X}B = 109°28'$ at the central anion in the AB_3 tetrahedron: then $\varrho = \sqrt{2/3} = 0.866$, and $u = 0.407$.

3.3 The Garnet ($H3_1/S1_4$) Structure

In garnets, $A_3B_2C_3X_{12}$, all the anions (X) are in A_2BC tetrahedra (Sect. 2.9.1 above). There are four independent bond distances from X to the corners of the A_2BC tetrahedra, and four structural parameters x, y, z and a. Again, using known bond lengths, the structural parameters can be calculated. (Note that this requires that the two A–X

19 Cf. also Ref. 88

bond lengths be equal, as they very nearly are in practice.) Thus if α, β, γ are l(A–X), l(B–X) and l(C–X) respectively, one has

$$x = 2(\gamma^2 - \alpha^2)/a^2 \tag{7}$$

$$z = 5/32 + (2\beta^2 - \alpha^2 - \gamma^2)/a^2 \tag{8}$$

$$y = (x + z)/2 \tag{9}$$

$$a = \beta/(x^2 + y^2 + z^2)^{1/2} \tag{10}$$

Solving these equations for $Y_3Fe_2Fe_3O_{12}$, taken as an example of a garnet with a well-refined structure[89], Shannon and Prewitt's table[84] for bond lengths yields the following parameters (observed[89] values are in parentheses):

a = 12.374 (12.376) Å ; x = − 0.0295 (− 0.0269) ,

y = 0.0599 (0.0581) , z = 0.1494 (0.1495) .

The agreement is typical but on reflection not surprising: indeed if the observed and calculated parameters differed significantly one should suspect the structure determination (as it would imply unusual bond lengths). It is well to recall too that in the early days of X-ray diffraction, when it was difficult to determine the coordinates of light atoms, they were often obtained by just this procedure.

The lattice parameters reported for garnets cover a wide range and it is interesting that the observed values agree very well with those calculated from Eqs. (7)–(10), as shown in Fig. 38.

3.4 *The Rutile (C4) and Zircon ($H0_3$, $S1_1$) Structures*

The rutile and zircon structures are considered together as they have related cation arrangements[90].

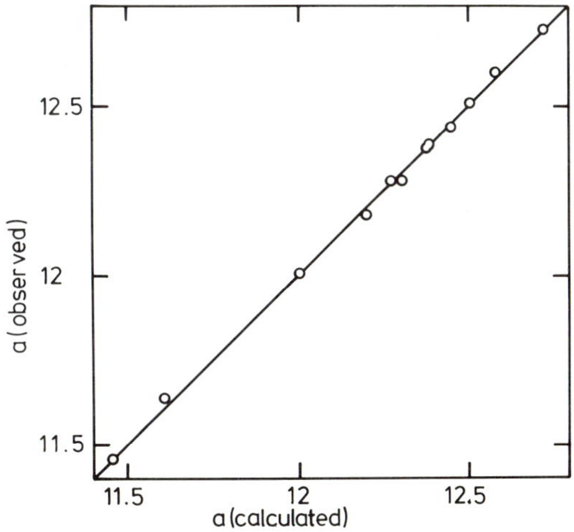

Fig. 38. A comparison of the observed and calculated (Sect. 3.3) unit cell parameters for various garnets

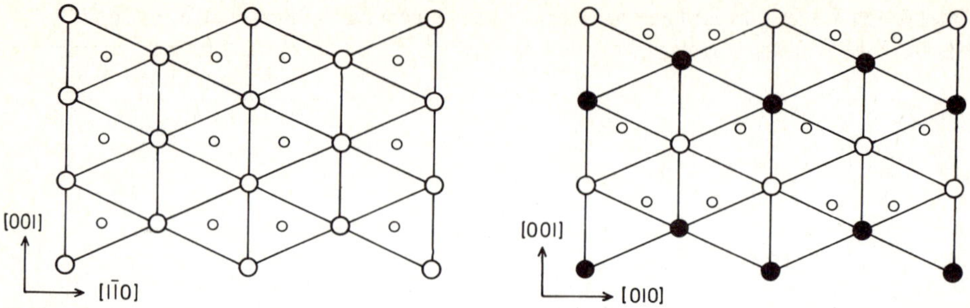

Fig. 39 a, b. A plane of (**a**) the rutile structure (*large circles* Ti, *small circles* O) and (**b**) the zircon structure (*large open circles* Zr, *large filled circles* Si, *small circles* O)

The cation arrangement of rutile is that of a body-centred tetragonal lattice ($c/a = 0.64$) such as has been found for the structure of β-Hg (for which $c/a = 0.71$). In this arrangement there are slightly distorted, close-packed (3^6) layers of Ti parallel to (110) as shown in Fig. 39 a. Oxygen atoms also lie in these layers, centering one half of the Ti_3 triangles. The triangles have one short edge, which corresponds to the short Ti...Ti distance required by the presence of four-membered TiOTiO rings (in conventional terms, edge-sharing TiO_6 octahedra).

The cation arrangement in zircon ($ZrSiO_4$) is essentially the same except that ordering of Zr and Si increases the size of the unit cell. Thus for rutile $a = 4.59$ Å, $c = 2.96$ Å and for zircon $a = 4.68 \times \sqrt{2}$ Å, $c = 3.01 \times 2$ Å. The layers are divided into equal numbers of $ZrSi_2$ and Zr_2Si triangles. Oxygen fills all the latter so that one now has $^{VIII}Zr\,^{IV}SiO_4$ instead of $^{VI}TiO_2$. The arrangement of atoms in a (100) plane of zircon [corresponding to a (110) plane of rutile] is shown in Fig. 39 b.

It has been shown earlier[9] that the parameters of the rutile structure can be very simply derived by setting the Ti–O bond lengths to the value expected for octahedral coordination (1.95 Å) and then allowing the structure to relax to maximum volume. There are three parameters: the lattice parameters a and c and a parameter x characterising the oxygen positions. The calculated values are (observed in parentheses)

$$a = 4.596\ (4.594)\ \text{Å}\ ,$$

$$c = 2.907\ (2.958)\ \text{Å}\ ,$$ [20]

$$x = 0.300\ (0.305)\ .$$

It is interesting and instructive to attempt to calculate the parameters of the zircon structure in the same way as was done for rutile. In addition to the lattice parameters a and c, there are two parameters required to specify the oxygen positions, which are 16(h) of $I4_1/amd$, viz. $0,x,z$ etc. Si atoms are in 4(b); 0,0,1/2 etc., and Zr in 4(a); 0,0,0 etc.

The oxygen atom at $0,x,z$ has one silicon neighbour at 0,0,1/2 with l(Si–O) = δ given by:

20 It is clear that the discrepancy between calculated and observed values of c is the result of ignoring strong Ti...Ti repulsions across the shared edges between TiO_6 octahedra (across the four-membered TiOTiO rings referred to above)

Table 4. Observed and calculated parameters and distances (in Å) in the zircon structure

	Observed	Calculated ($\alpha = \beta$)
a	6.607	6.674
c	5.982	5.905
x	0.184	0.177
z	0.320	0.312
α	2.268	2.188
β	2.131	2.188
O...O(1)	2.429	2.360
O...O(2)	2.753	2.777
Zr...Si	2.991	2.952
Zr...Zr	3.626	3.649

$$\delta^2 = x^2a^2 + (1/2 - z)^2c^2 \tag{11}$$

It also has two Zr atom neighbours at 0,0,0 and 0,1/2,1/4 with two Zr–O distances, α and β, given by:

$$\alpha^2 = x^2a^2 + z^2c^2 \tag{12}$$

$$\beta^2 = (1/2 - x)^2a^2 + (1/4 - z)^2c^2 \tag{13}$$

Proceeding as with rutile, with $\delta = 1.62$ Å and $\alpha = \beta = 2.19$ Å, and then maximising the volume, one obtains the parameters listed in Table 4. They are in fair agreement with the observed parameters[91], which are also listed.

In the observed structure the two Zr–O distances are significantly different. The long bonds are in ZrOSiO rings allowing the short Zr...Si and O...O distances to increase, and it might be supposed that the occurrence of unequal Zr–O bonds arises from these non-bonded repulsions.

3.5 The Bixbyite ($D5_3$) Structure

The bixbyite (C-type M_2O_3) structure is cubic, space group $Ia3$ with M(1) in 8(a) 0,0,0 etc., M(2) in 24(d) x,0,1/4 etc., and O in 48(e) x,y,z etc. The structure therefore has five free parameters.

It is interesting to recall that in their original determination of the structure (in 1930) Pauling and Shapell[92] determined only the unit-cell edge a and x for M(2). The oxygen parameters were then deduced by requiring that the three independent M(2)–O distances were equal to that of M(1)–O. This resulted in oxygen parameters very close to those found in subsequent accurate determinations.

In a similar vein it was later shown[9] that all the internal parameters [x for M(2) and x,y,z for 0] could be determined by maximising the volume subject to the constraint of all four M–O distances being kept equal and constant. Finally, taking the M–O distance from tables[84], the lattice parameter could be calculated. Table 5 lists parameters for the real, the maximum volume, and the "ideal" [metal atoms as in **Cu_3Au**, O centering the

Table 5. The real structure of Sc_2O_3 compared with the "ideal" structure and the maximum volume one (calculated with Sc–O = 2.128 Å)

	Ideal	Real	Maximum volume
x, M(2)	0	−0.035	−0.027
x	0.375	0.393	0.384
y	0.125	0.152	0.145
z	0.375	0.384	0.381
a	9.827	9.847	9.919
d(M...M)	3.474	3.25–3.74	3.32–3.71
d(O...O)	2.46	2.58–2.68	2.63–2.66

$M(1)M(2)_3$ tetrahedra] structures. It is worth pointing out that the real and ideal structures differ: in the former the O...O distances are greater and the M...M distances smaller so that, in this instance, it is anion... anion repulsions that provide the driving force for departure from the "ideal". This is not surprising as in the ideal structure the metal-atom array is the most efficient packing of c.c.p. whereas the anion array is the very open one of three-quarters of the sites of a simple-cubic array.

3.6 The Scheelite (HO_4) Structure

The structure of scheelite is tetragonal, symmetry $I4_1/a$. Ca atoms are in 4(b) 0,0,1/2 etc., and W atoms in 4(a) 0,0,0 etc. The axial ratio is c/a = 2.09: if it were 2.00, Ca and W taken together would lie exactly on the nodes of a face-centered-cubic lattice, with all the tetrahedra of the type Ca_2W_2[21]. Oxygen atoms are in 16(f) x,y,z etc.

For the oxygen atoms to centre the tetrahedra $x = y = 1/4$, $z = 1/8$. The symmetry would now be $I4_1/amd$, with oxygen in 16(f) of that group: x,1/4,1/8 with $x = 1/4$. In fact the oxygen atoms are in Ca_2W triangles so that the metal coodination is CaO_8 and WO_4. The W–O distance α is given by

$$\alpha^2 = (x^2 + y^2)a^2 + z^2c^2 , \tag{14}$$

and the Ca–O distances β and δ by

$$\beta^2 = (1/2 - x)^2a^2 + (1/2 - y)^2a^2 + z^2c^2 \tag{15}$$

$$\delta^2 = (1/2 - x)^2a^2 + y^2a^2 + (1/4 - z)^2c^2 \tag{16}$$

Proceeding as with other structures described earlier, i.e. taking for α and β (assumed equal to δ) the expected bond lengths 1.79 Å and 2.47 Å respectively, we can attempt to calculate the structural parameters. As there are five parameters and only three bond lengths we again simulate the non-bonded repulsions by maximising the crystal volume subject to the constraints, Eqs. (14–16). The result is compared with the observed structure in Table 6. It can be seen that the agreement is quite good.

21 If c/a were equal to $\sqrt{2}$ the cation array would be that of **NaTl,** a superstructure of body-centred cubic, and if $c/a \approx 1$ the cation array would be that of zircon ($ZrSiO_4$) and of KH_2PO_4

Table 6. Calculated and observed parameters in scheelite ($CaWO_4$)

	Calculated	Observed
a	5.370	5.243
c	11.235	11.376
x	0.270	0.241
y	0.130	0.150
z	0.070	0.086

It is worth noting an important difference between this structure and that of zircon. In scheelite the OCa_2W triangles share only the Ca_2 edges, so that the rings of four atoms are CaOCaO. For this reason the two Ca–O bonds are of approximately equal length (2.44 and 2.48 Å) in the observed structure. The short Ca . . . Ca distance is 3.86 Å and the O . . . O distance is 2.75 Å in this ring. (The analogous rings in zircon are of two types, ZrOSiO as well as ZrOZrO.

3.7 *The Calcite* (GO_1) *Structure*

Of the structures with stoichiometry ABO_3 that may be derived from that of perovskite (Sect. 2.2), the calcite structure is the simplest in the sense that it requires the fewest parameters to specify it. We earlier described[25)] this structure in terms of the parameters obtained with regular BO_6 octahedra that are rotated (tilted) about their $\overline{3}$ axes from the positions they have in (cubic) perovskite. The coordination of A goes from 12 in the cubic structure to 3 in the rhombohedral calcite structure.

In calcite, however, the CaO_6 octahedra are not regular, and it is again revealing to attempt to calculate the structure parameters using the methods of earlier sections (i.e. using known bond lengths and maximising the volume).

The symmetry of the structure is $R\overline{3}c$ with Ca in 6(*b*) 0,0,0 etc., C in 6(*a*) 0,0,1/4 etc. and O in 18(*e*) *x*,0,1/4 etc. (hexagonal unit cell). The oxygen atoms are each bonded to two equidistant Ca atoms and one C atom. The C–O bond length is

$$\alpha = ax \tag{17}$$

and the Ca–O bond length is given by

$$\beta^2 = c^2/144 + a^2(1 - 3x + 3x^2)/3 \; . \tag{18}$$

The volume, $\sqrt{3}\,a^2c/2$, will be a maximum for

$$a = 5\alpha/4 + (2\beta^2 - 7\alpha^2/16)^{1/2} \tag{19}$$

$$x = \alpha/a \tag{20}$$

$$c = 12[\beta^2 - (1 - 3x + 3x^2)a^2/3]^{1/2} \; . \tag{21}$$

With $\alpha = 1.28$ Å, $\beta = 2.35$ Å this yields (observed values in parentheses)

$$a = 4.814\ (4.990)\ \text{Å}$$

$$c/a = 3.80\ (3.41)$$

$$x = 0.266\ (0.257)\ .$$

The agreement is fair, showing again that using known bond lengths and the simple idea that non-bonded repulsions can be simulated by maximising the volume gives a good first approximation to a crystal structure. Also, by examining the difference between the observed and calculated distances around the oxygen atom, we get further insight into the nature of the non-bonded forces. In the $OCCa_2$ group (which is not quite planar) the calculated (observed) distances are:

$$d(\text{C}\ldots\text{Ca}) = 3.170\ (3.211)\ ,$$

$$d(\text{Ca}\ldots\text{Ca}) = 4.126\ (4.043)\ \text{Å}\ .$$

There is also a short O . . . O distance in the structure (2.221 Å) in CaOCO rings, but this is determined solely by the C–O bond length (i.e. it is the edge of an equilateral CO_3 triangle). The main difference from the calculated structure is that the C . . . Ca distance is increased in the observed structure at the expense of a decreased Ca . . . Ca distance. (Compare also the observed and calculated c/a values above.) This suggests that in the calculated structure the C . . . Ca repulsions would be greater than the Ca . . . Ca repulsions.

Note however that the maximum volume is only 0.05% larger than the actual one.

The CCa packing is sometimes described as a **NaCl** (B 1) arrangement, which it would be if $c/a = 4.90$. Our description as derived from perovskite emphasises a CCa packing derived from **CuZn** (B 2). This latter would require $c/a = 2.45$ which is slightly closer to the observed value of $c/a = 3.41$.

The structure is also described as based on a hexagonal close packing of oxygen. However, O . . . O distances within a layer range from 3.261 Å to 2.221 Å. The latter distance might be compared to twice the conventional ionic radius of oxygen, viz. 2.6–2.8 Å.

4 The Effect of Pressure on Structure

The crystal property perhaps most sensitive to structure is density and, conversely, the increase in density of a crystal with pressure is accompanied by significant structural changes. The latter can be of two types – either a smooth variation of the free parameters of the structure with pressure, or a first-order transition to a new structure type.

Good data pertaining to the first type of structural change are rather scarce because of the difficulty of making accurate structure determinations at high pressure. However, where they exist they are informative about the relative importance of the various atom ... atom near-neighbour interactions. In particular, of near neighbours that are repelling each other, it might be expected that those repelling most strongly will move the least towards each other, whereas those more weakly repelling will move rather more. We give just one example which illustrates this point.

Forsterite (Mg_2SiO_4, olivine type) contains four MOMO rings with short M ... M and O ... O distances: two are MgOSiO, two are MgOMgO. When it is compressed from 1 atmos (~ 0 kbar) to 20 kbar the parameter changes[93)] reveal the following proportional decreases in these short second-nearest neighbour distances: Mg ... Si = 0.4_0%, Mg ... Mg = 0.6_7%, O ... O = 2.1_6%. These figures clearly suggest that O ... O repulsions are less important than those of Mg ... Mg which, in turn, are somewhat smaller than those of Mg ... Si; i.e. that cation ... cation repulsions are greater than anion ... anion repulsions, and that the Mg ... Si forces are the more important of these. This conclusion is reinforced by other observations, to be described below (Sect. 4.1.1).

4.1 Phase Transitions and Packing Density

In a solid, the occurrence of a phase transition with applied pressure can also provide valuable insight into the relative importance of the various interatomic repulsions in the crystal. In most cases the strongest repulsive component is that between atoms that are bonded together (usually nearest neighbours), and the effect of pressure is to produce a new structure of lower volume in which these nearest neighbours are *further apart,* and (usually) next-nearest neighbours are closer together. Necessarily then, there is an increase in primary coordination number (of both cations and anions, of course). Thus, consider the prototypical sphalerite (B 2) → rock salt (B 1) phase transition for composition MX. The B 1 phase is the denser and so, because the M and X arrays are identical in both structures, all M ... M and X ... X distances are shorter. However, by virtue of the fact that M moves from a tetrahedron of X to an octahedron of X (and similarly for X in M) the shortest M–X distances are longer.

A second, less common kind of pressure-induced phase transition (but in one instance at least, of very great importance) is that in which there is no change in primary coordination number and, to a good approximation, no change in nearest-neighbour bond lengths but, nevertheless, a substantial decrease in volume. We consider next, two examples of this latter type of phase transition.

4.1.1 Olivine → Spinel

These two structures have been described above (Sect. 2.6.3 and Sects. 2.8.1, 3.2) and are depicted in Figs. 16 and 29. We have already suggested (Sect. 4) that in forsterite (α-Mg_2SiO_4, olivine type) the repulsive forces between O ... O ($d \approx 2.6$ Å) are less than those between Mg ... Mg ($d \approx 3.0$ Å) which, in turn, are less than those between Mg ... Si ($d \approx 2.7$ Å). One would therefore expect that in the phase transition α-Mg_2SiO_4 (olivine) → γ-Mg_2SiO_4 (spinel) the principal result would be an increase in the last of these distances, in order to relieve the largest non-bonded repulsions[95)]. And this is indeed the case: the mean non-bonded (next-nearest neighbour) distances are as shown

Table 7. Mean shortest non-bonded distances (in Å) in the α- and γ-polymorphs (olivine and spinel forms) of (a) Mg_2SiO_4[93, 94], (b) Fe_2SiO_4[62]

	α-form	γ-form	Difference, Δ(γ − α)
a) Mg_2SiO_4			
Mg...Si	3.26_5 (2.69_2–3.28_3)	3.36_6	$+0.10_1$
Mg...Mg	3.11_5 (2.99_3–3.86_2)	2.87_0	-0.24_5
O...O	2.91_2 (2.55_8–3.39_3)	2.85_2 (2.70_3–3.00_0)	-0.06_0
b) Fe_2SiO_4			
Fe...Si	3.14_7 (2.78_5–3.31_8)	3.41_4	$+0.26_7$
Fe...Fe	3.25_3 (3.05_0–3.30_5)	2.91_1	-0.34_2
O...O	2.97_6 (2.57_3–3.27_8)	2.912 (2.699–3.124)	−0.064

in Table 7 a. (Where there is a range of d values, these are given in parentheses.) Similar data for Fe_2SiO_4 are given in Table 7 b.

Not unexpectedly, the AB_n coordination number increases from $n = 9$ to $n = 12$ in this transformation. As a result (cf. the discussion of B 2 → B 1 in Sect. 4.1), the A...B distance increases, while the B...B distance decreases. And, since the primary coordination numbers of the anions and both types of cations remain unchanged by the transformation, the A–O and B–O distances are hardly altered, as are the O...O distances. (The oxygens appear to be irrelevant in the transition!)

The changes in non-bonded distances are somewhat larger for Fe_2SiO_4 than for Mg_2SiO_4. It may be relevant that the transformation pressure is larger for the latter: 38 kbar and 155 kbar respectively. The structures were, of course, determined at 1 atm. pressure; the distances would be different at the equilibrium pressure.

The results suggest that, if A...B repulsions dominate, then the spinel form is stable with respect to the olivine form; and that if B...B repulsions dominate, the reverse is true. (The relative "hardness" of A and B is important, i.e. the slopes of their curves of potential energy as a function of separation.)

4.1.2 Zircon → Scheelite

In some respects this high-pressure transformation is similar to that of olivine → spinel: there is a large volume decrease (~11%; cf. ~10% for olivine → spinel), but no change in primary coordination number (OZr_2Si, ZrO_8, SiO_4). We have already considered the two structure types involved (Sect. 3.4 and Sect. 3.6).

In zircon (the normal pressure form of $ZrSiO_4$) there are ZrOSiO rings in which the Zr...Si distance is rather short, 2.99_1 Å; and (in Sect. 3.4) we have noted that this probably indicates a strong non-bonded repulsion between these two atoms, so that the associated Zr–O bonds are lengthened: they are 2.26_8 Å compared with 2.13_1 Å for the shorter of the two Zr–O bond types. On the other hand, in the high-pressure (scheelite) polymorph of $ZrSiO_4$ these repulsions are eliminated: there are no ZrOSiO rings – only ZrOZrO. And the Zr...Si distance is far too long (3.34_7 Å) for any significant Zr...Si repulsion. It seems clear then, that the driving force for the transition under pressure is the reduction in the increasing repulsion energy due to non-bonded contacts between the two types of cations – exactly as in the olivine → spinel transformation.

Finally, it seems to us significant that these two structures are closely related to the rutile and fluorite types, respectively, and that the volume change in the transformation is very similar to that for rutile → fluorite in oxides.

5 Concluding Remarks

It seems necessary to point out the essence of what we are saying – and what we are not saying – in this paper. There are two main levels of approach.

First, there is the purely descriptive, geometrical, phenomenological point that in many crystals – *but not, by any means, in all* – a quite superficial examination of the structure reveals that the cation array is quite regular and, furthermore, that the same arrangements of cations are frequently those observed in the structures of metal alloys (including borides, silicides etc.) or of plausible (though unknown) alloy structure types. In such cases the oxygens (or other small anions) are located in sites that are perfectly logical and appropriate – most frequently tetrahedra – even when the anion array alone defies simple description, and is by no means regular. In these latter cases we believe that the cation-emphasised description is often the first and only simple description of such structures. (Examples are some sulphates, apatite etc.) It is also often the case that a given cation array can, in different structures, contain a different number and/or a different arrangement of anions, suggesting that the cation arrangement has some significance in its own right, and at least some semblance of "permanency", e.g. bixbyite/pyrochlore, garnet/Ag_3PO_4 etc. This conclusion is supported by the many examples of structure/antistructure relations (complete or partial) adduced in Tables 1 and 2.

At the second level the question arises as to why such simple and well-known cation arrangements appear in oxides. It seems to us plausible that this is a consequence of important non-bonded repulsions between the cations. Our belief is that these are frequently greater than anion . . . anion repulsions. And certainly this is consistent with the observed regularity of the cation arrays – which is often greater than that of the anion arrays. (Good examples are provided by various olivines and humites, an extreme case being that of the chondrodite type $2\,Cd_2SiO_4 \cdot Cd(OH)_2$[96], which is shown in Fig. 25.)

We repeat that this is by no means the first time that attention has been drawn to the nature of the cation arrangements in various oxides etc. We have, for example, cited the case of Geller's description of the cation array in garnet as A 15 type[67]. And Gorter (and others have) certainly pointed out that the cation array in spinel is C 15 ($MgCu_2$) type[97]. Nevertheless this has usually been only for specific instances; i.e. in such cases there has been no suggestion that this was a useful principle. The implication, if any, was that it was an interesting, perhaps amusing, curiosity.

However, in one or two cases it was stated or implied that it was a more general phenomenon of considerable importance to the crystal chemist. (But we have discovered these only in the last stages of writing this paper.) Perhaps the most forthright assertion is by Lebedev[44, 98], who has published a great deal on this topic in the Russian literature. He makes the point that many isostructural pairs of compounds which, in terms of the idea that oxygens are close-, or at least densely, packed, should have the same molar volumes and very different densities in fact show the inverse: i.e. very different molar volumes and almost equal densities. His explanation is that cations are large and anions small, and it is the former that are densely packed. We differ from him in believing that, in oxides, the spacing of densely-packed cations (which are next-nearest neighbours) depends on their non-bonded rather than their ionic radii.

The ideas in this paper concerning bond lengths are close in spirit to the those expressed by Zachariasen[99–102] who[100] has, in particular, broadened Pauling's bond strength rules to cover both (articulated) anion- and cation-centred polyhedral arrays. We believe that his approach to bond length/bond strength relations[101, 102] is of especial power and relevance to understanding the dimensional details of simple inorganic structures. (Cf., for example, our treatment of **A-La_2O_3** in Sect. 2.1.2.2.) It is far more rational, and more accurate and widely-applicable, than the conventional approach to structures in terms of dense anion packing and determination of the sites occupied by means of the radius ratio rule, with its exclusive emphasis on cation-centred polyhedra of anions.

We should not omit mention of the paper "Oxygen as the coordination centre in crystal structures" by Bergerhoff and Paeslack[12] (and subsequently by Caro[13]) and the use of "an *artificial* parameter described as a force constant between A and B atoms" (in AB_2X_4 structures) by Kugimiya and Steinfink[103] (our italics). However, in neither of these cases is there any analysis of the geometry of the cation array.

Finally, we are *not* saying that anion ... anion repulsion is always negligible, and that cation ... cation repulsion alone determines the nature of an oxide structure; see, for example, Sect. 3.5. (But neither would we agree with the reverse statement.) However, for compounds with stoichiometric ratios anion/cation ≤ 2 and with first row anions, cation ... cation repulsion is likely to dominate.

The striking fact about structures is Nature's parsimony in selecting appropriate geometrical patterns. This has long been recognised in anion arrays. We point out here that it is equally true of cation arrays.

Acknowledgements. This work was supported by a grant from the U.S. National Science Foundation (DMR 8119061). We would also like to thank the Arizona State University for assistance in the form of a Visiting Professorship (to BGH), Mrs. Jane Sutton for a great deal of typing and patience, and Mrs. Manea Southwell for photocopying the many drawings.

6 References

1. O'Keeffe, M., Hyde, B. G.: Trans. Amer. Cryst. Assoc., *15,* 65 (1979)
2. O'Keeffe, M., Hyde, B. G.: Chapter 10 in: Structure and Bonding in Crystals (eds. O'Keeffe, M., Navrotsky, A.) New York, Wiley 1981
3. Schubert, K.: Kristallstrukturen Zweikomponentiger Phasen, Berlin, Springer-Verlag 1964
4. Wells, A. F.: Structural Inorganic Chemistry, 4th ed. Oxford, Clarendon Press 1975
5. Pearson, W. B.: The Crystal Chemistry and Physics of Metals and Alloys, New York, Wiley 1972
6. Hyde, B. G., Andersson, S., Bakker, M., Plug, C. M., O'Keeffe, M.: Prog. Solid State Chem. *12,* 273 (1979)

7. Nyman, H., Andersson, S., Hyde, B. G., O'Keeffe, M.: J. Solid State Chem. *26,* 123 (1979)
8. Andersson, S.: Chapter 24 in: Structure and Bonding in Crystals (eds. O'Keeffe, M., Navrotsky, A.) New York, Wiley 1981
9. O'Keeffe, M.: Acta Cryst. *A33,* 924 (1977)
10. O'Keeffe, M., Bovin, J.-O.: Amer. Miner. *63,* 180 (1978)
11. Hyde, B. G.: Acta Cryst. *A27,* 617 (1971)
12. Bergerhoff, G., Paeslack, J.: Z. Kristallogr. *126,* 1 (1968)
13. Caro, P.: Solid State Chemistry, N.B.S. Spec. Publ. 364 (eds. Roth, R. S., Schneider, S. J.) Washington: U.S. Government Printing Office 1972, p. 367
14. O'Keeffe, M., Hyde, B. G.: Acta Cryst. *B34,* 3519 (1978)
15. Bowman, A. L., Arnold, G. P.: ibid. *B27,* 243 (1971)
16. Sibieude, F.: J. Solid State Chem. *7,* 7 (1973)
17. Sibieude, F., Schiffmacher, G., Caro, P.: ibid. *23,* 361 (1978)
18. O'Keeffe, M., Hyde, B. G.: Phil. Mag. *33,* 219 (1976)
19. Bevan, D. J. M., Summerville, E.: Chapter 28 in Handbook on the Physics and Chemistry of Rare Earths *3,* (eds. Gschneidner, K. A., Eyring, L.) Amsterdam, North Holland 1979
20. Hyde, B. G., Bevan, D. J. M., Eyring, L.: Phil. Trans. Roy. Soc. (Lond.) *A259,* 583 (1966)
21. Van der Walt, C. M., Louw, H. N. J.: Acta Met. *16,* 777 (1968)
22. Rudy, E.: Met. Trans. *1,* 1249 (1970)
23. Glazer, A. M.: Acta Cryst. *B28,* 3384 (1972)
24. O'Keeffe, M., Bovin, J.-O., Hyde, B. G.: Phys. Chem. Minerals *4,* 299 (1979)
25. O'Keeffe, M., Hyde, B. G.: Acta Cryst. *B33,* 3802 (1977)
26. O'Keeffe, M.: Chapter 13 in Structure and Bonding in Crystals (eds. O'Keeffe, M., Navrotsky, A.) New York, Wiley 1981
27. Bragg, W. L., Claringbull, G. F.: Crystal Structures of Minerals. London, Bell 1965
28. Wyckoff, R. W. G.: Crystal Structures *3,* 46. New York, Interscience 1965
29. Wyckoff, R. W. G.: ibid. *4,* 545. New York, Interscience 1965
30. Hill, R. J.: Canad. Miner. *15,* 522 (1977)
31. Flörke, O. W.: Naturwissenschaften *39,* 478 (1952)
32. Nowotny, H., Szekely, G.: Monatsh. Chem. *83,* 568 (1952)
33. Wyckoff, R. W. G.: Crystal Structures *3,* 56. New York, Interscience 1965
34. Steinfink, H., Brunton, G. C.: Acta Cryst. *B24,* 807 (1968)
35. Bats, J. W.: ibid. *B34,* 1679 (1978)
36. Ramachandran, G. M., Lonappan, M. A.: ibid. *10,* 281 (1957)
37. Wyckoff, R. W. G.: Crystal Structures *3,* 101–102. New York, Interscience 1965
38. McGinnety, J. A.: Acta Cryst. *B28,* 2845 (1972)
39. Jost, K. H., Ziemer, B., Seydel, R.: ibid. *B33,* 1696 (1977)
40. Smith, D. K., Majumdar, A., Ordway, F.: ibid. *18,* 787 (1965)
41. Barnes, P., Fentiman, C. H., Jeffery, J. W.: ibid. *A36,* 353 (1980)
42. Donohue, J.: The Structures of the Elements, New York, Wiley 1974
43. Wyckoff, R. W. G.: Crystal Structures *1,* 241. New York, Interscience 1965
44. Lebedev, V. I.: Internat. Geol. Rev. *14,* 543 (1972)
45. Miyake, M., Morikawa, H., Iwai, S.-I.: Acta Cryst. *B36,* 532 (1980)
46. Hanic, F., Handlovic, M., Kurdova, K., Majling, J.: J. Crystall. Spectrosc. Res. *12,* 99 (1982)
47. Groves, G. W.: J. Mater. Sci. *18,* 1615 (1983)
48. Czaya, R.: Acta Cryst. *B27,* 848 (1971)
49. Johnson, V.: Inorg. Chem. *14,* 1117 (1975)
50. Jeitschko, W.: Acta Cryst. *B31,* 1187 (1975)
51. Wenk, H.-R., Raymond, K. N.: Z. Kristallogr. *137,* 86 (1973)
52. Hyde, B. G., Bagshaw, A. N., Andersson, S., O'Keeffe, M.: Ann. Rev. Mater. Sci. *4,* 43 (1974)
53. Andersson, S., Hyde, B. G.: J. Solid State Chem. *9,* 92 (1974)
54. Gibbs, G. V., Ribbe, P. H.: Amer. Miner. *54,* 376 (1969)
55. Gibbs, G. V., Ribbe, P. H., Anderson, C. P.: ibid. *55,* 1182 (1970)
56. Ribbe, P. H., Gibbs, G. V.: ibid. *56,* 1155 (1971)
57. Robinson, K., Gibbs, G. V., Ribbe, P. H.: ibid. *58,* 43 (1973)
58. White, T. J., Hyde, B. G.: Acta Cryst. *B39,* 10 (1983)

59. Dunn, P. J., Peacor, D. R., Simmons, W. B., Essene, E. J.: private communication (1983)
60. White, T. J., Hyde, B. G.: Amer. Miner. *68,* 1009 (1983)
61. Brookin, A. A., Nikishova, L. V.: Sov. Phys. Crystallogr. *20,* 252 (1975)
62. Marumo, F., Isobe, M., Akimoto, S.: Acta Cryst. *B33,* 713 (1977)
63. Daniel, F., Moret, J., Philippot, E., Maurin, M.: J. Solid State Chem. *22,* 113 (1977)
64. Novak, G. A., Gibbs, G. V.: Amer. Miner. *56,* 791 (1971)
65. Hawthorne, F. C.: J. Solid State Chem. *37,* 157 (1981)
66. O'Keeffe, M., Andersson, S.: Acta Cryst. *A33,* 924 (1977)
67. Geller, S.: Z. Kristallogr. *125,* 1 (1967)
68. Schwartz, K. B., Parise, J. B., Prewitt, C. T., Shannon, R. D.: Acta Cryst. *B38,* 2109 (1982)
69. Arem, J. E., Burnham, C. W.: Amer. Miner. *54,* 1546 (1969)
70. Ito, J., Arem, J. E.: ibid. *55,* 880 (1970)
71. Rucklidge, J. C., Kocman, V., Whitlow, S. H., Gabe, E. J.: Canad. Miner. *13,* 15 (1975)
72. Warren, B. E., Modell, D. I.: Z. Kristallogr. *78,* 422 (1931)
73. Andersson, S.: J. Solid State Chem. *23,* 191 (1978)
74. McConnell, D., Verhoek, F. H.: J. Chem. Ed. *40,* 512 (1963)
75. Sudarsanan, K., Young, R. A.: Acta Cryst. *B25,* 1534 (1969)
76. Nyman, H., Andersson, S.: ibid. *A35,* 580 (1979)
77. Sudarsanan, K., Young, R. A.: ibid. *B34,* 1401 (1978)
78. Besse, J.-P., Band, G., Levasseur, G., Chevalier, R.: ibid. *B35,* 1756 (1979)
79. Besse, J.-P., Band, G., Chevalier, R., Zarembowitch, J.: Mater. Res. Bull. *15,* 1255 (1980)
80. Band, G., Besse, J.-P., Capestan, M., Sueur, G., Chevalier, R.: Ann. Chim. *5,* 575 (1980)
81. Newnham, R. E., Wolfe, R. W., Darlington, C. N. W.: J. Solid State Chem. *6,* 378 (1973)
82. Otto, H. H.: Z. Kristallogr. *160,* 93 (1982)
83. Smith, J. V.: Amer. Miner. *63,* 960 (1978)
84. Shannon, R. D., Prewitt, C. T.: Acta Cryst. *B25,* 925 (1969)
85. Barker, W. W., Graham, J., Knop, O., Brisse, F.: in: The Chemistry of Extended Defects in Nonmetallic Crystals (eds. Eyring, L., O'Keeffe, M.) Amsterdam, North Holland 1970, p. 198
86. Brown, I. D.: Chapter 14 in Structure and Bonding in Crystals (eds. O'Keeffe, M., Navrotsky, A.) New York, Wiley 1981
87. Becker, W.-J., Will, G.: Z. Kristallogr. *131,* 278 (1970)
88. Hill, R. J., Craig, J. R., Gibbs, G. V.: Phys. Chem. Minerals *4,* 317 (1979)
89. Batt, A., Post, B.: Acta Cryst. *15,* 1268 (1962)
90. O'Keeffe, M., Hyde, B. G.: J. Solid State Chem. *44,* 24 (1962)
91. Robinson, K., Gibbs, G. V., Ribbe, P. H.: Amer. Miner. *56,* 782 (1971)
92. Pauling, L., Shapell, M. D.: Z. Kristallogr. *75,* 128 (1930)
93. Hazen, R. M.: Amer. Miner. *61,* 1280 (1976)
94. Sasaki, S., Prewitt, C. T., Sato, Y., Ito, E.: J. Geophys. Res. *87,* 7829 (1982)
95. O'Keeffe, M., Hyde, B. G.: Nature *293,* 727 (1981)
96. Egorov-Tismenko, Yu. K., Duneko, I. P., Simonov, M. A., Belov, N. V.: Sov. Phys. Crystallogr. *16,* 1031 (1972)
97. Gorter, E. W.: Philips Res. Rep. *9,* 295 (1954)
98. Lebedev, V. I.: Sov. Phys. Crystallogr. *17,* 33 (1972)
99. Zachariasen, W. H.: Acta Cryst. *7,* 795 (1954)
100. Zachariasen, W. H.: ibid. *16,* 385 (1963)
101. Zachariasen, W. H.: J. Less-Common Metals *62,* 1 (1978)
102. Zachariasen, W. H.: ibid. *69,* 369 (1980)
103. Kugimiya, K., Steinfink, H.: Inorg. Chem. *7,* 1762 (1968)
104. Subramanian, M. A., Aravamudan, G., Subbarao, G. V.: Progr. Solid State Chem. *15,* 55 (1983)
105. Quoted in Mackay, A. L.: The Harvest of a Quiet Eye (ed. Ebison, M.), Bristol, Institute of Physics 1977
106. Wondratschek, H., Merker, L., Schubert, K.: Z. Kristallogr. *120,* 393; 478 (1964)
107. Merker, L., Wondratschek, H.: Z. Kristallogr. *125,* 1 (1967)

Redox Reactions of Sulphur-Containing Amino-Acid Residues in Proteins and Metalloproteins, an XPS Study

Ulrich Weser

Anorganische Biochemie, Physiologisch-chemisches Institut der Universität Tübingen, Hoppe-Seyler-Str. 1, 7400 Tübingen 1, Federal Republic of Germany

Sulphur containing amino acid residues are found among the structural and functional constituents of many proteins. The ease of oxidising sulphur in these biologically important polymers is well known. At present, there is no simple and fast technique available to monitor the respective oxidation states using chemical reactions. X-ray photoelectron spectroscopy has proven to be both convenient and efficient to follow these electron transport reactions. In the present survey a choice of some sulphur and selenium bearing proteins and metalloproteins will be examined upon excessive irradiation, treatment with hydrogen peroxide and cleavage of metal sulphur bonding. It will also be shown that the use of this technique in general will allow a quick decision on the rate of deterioration of a protein. The limitations of this spectrochemical approach will be debated.

1 Introduction

Sulphur and transition metals enjoy both a marked distribution and importance in biological electron transport reactions. There are homogeneous metal thiolate proteins including iron-sulphur and iron-molybdenum-sulphur proteins. Mixed ligands are found among the haem sulphur proteins where cysteine or methionine belong to the axial ligands of the haem moiety. Histidine, cysteine and/or methionine are the known sulphur bearing ligands forming the type I copper binding sites of many blue oxidases. Apart from the above listed electron transport proteins there are homogeneous Cu, Zn and Cd-thiolate proteins[1–3]. Their possible function is assigned to metal transportation[4, 5].

Unlike the specifically bound metals which participate in reversible redox reactions extraneously bound transition metals are well able to induce uncontrolled oxidation of sulphur containing ligands[2, 6]. The so called metal driven Haber Weiss reaction is responsible for the formation of oxidised sulphur species leading to sulphonic acid. For example:

$$RS^- + Cu(I) \rightleftharpoons \frac{1}{n}[Cu(I)SR]_n \xrightleftharpoons{+O_2} \frac{1}{n}[Cu(I)O_2SR]_n \rightleftharpoons Cu(II) + RS^- + O_2^{-}\bullet \qquad (1)$$

$$RS^- + Cu(II) \rightleftharpoons \frac{1}{n}[Cu(I)\bullet SR]_n \rightleftharpoons (RSSR)_{1/2}Cu(I) \qquad (2)$$

$$(RSSR)_{1/2}Cu(I) \rightleftharpoons \frac{1}{2}(RSSR) + Cu(I) \qquad (3)$$

$$Cu(I) + \bullet O_2^- \rightarrow Cu^{2+} + O^{2-} + \bullet O^- \qquad (4)$$

$$\bullet O^- + RS^- \xrightarrow{O_2} RSO_3^- \qquad (5)$$

The $\bullet O^-$ radical is well able to oxidise all sorts of sulphur species up to S(VI).

Excited oxygen species originating from biochemical reactions other than the direct redox cycles may affect intact sulphur containing amino acid residues. Irradiation damage must also be listed to the many possibilities in catabolizing these residues. It is needless to mention that the age dependent deterioration of proteins must be seen in relation to the same oxygen radical induced breakdown of vital sulphur containing amino acids. A suitable method to monitor the differently oxidised sulphur species was the use of X-ray photoelectron spectroscopy. This technique has proven to be rather elegant. Its use was extended on measurements of different oxidation states of selenium, an even more sensitive chalcogen. Selenium belongs to the biochemically active and essential elements and is also covalently bound in proteins. For example, the redox cycle of selenium in the reaction of glutathione peroxidase was shown to involve a selenol and Se(IV)[7–9].

2 Experimental

X-ray photoelectron spectra were run on a Varian V-IEE-15 spectrometer equipped with an on-line 8 K computer (620 L)[10,11]. Alternatively a Leybold LHS-10 unit was employed. In all experiments a magnesium anode was used, which is known to emit mainly Mg $K\alpha_{1,2}$ X-radiation. Calibration of the binding energies was achieved by employing the carbon 1 s line, which was attributed to aliphatic hydrocarbons with an arbitrarly fixed binding energy of 284 eV[7,8]. This reference line was measured each time a recording of the spectral regions of a particular sample had been made. Total scanning time was minimized to avoid deterioration of samples and was usually less than 15 min at an X-ray energy of 1100 W. As soon as the vacuum in the analyzer chamber reached 1 μTorr cooling of the samples to − 150 °C became necessary for the same reason. Solid samples were dusted onto a one-sided adhesive cellotape (Scotch 3 M Co) which covered the cylindrical sample holder. Liquids were applied on a specially pretreated aluminium sample holder. The pretreatment of this aluminium holder was carried out according to the procedure of Adams and Bancroft[12]. A newly worked, cylindrical aluminium sample holder, which fitted the Varian sample rod, was mechanically engraved (15 annular engravings per cm). Cleaning was carried out by lightly etching in 3 M HCl for 2–3 min until the surface of the sample holder was wetted readily by water. Etched aluminium cylinders were washed with water and acetone and dried at 70 °C. The sample holder treated in this way showed no signals in the S 2p binding-energy region[7]. Cytochrome *c* and myoglobin was from Sigma, München. Papain was from Boehringer, Mannheim, Cu-thionein was isolated from yeast[13–15], Cd-, Zn-thionein from rats[16], Cu_2Zn_2-superoxide dismutase was from bovine erythrocytes[17] and plastocyanin was prepared from parsley[18].

3 Inorganic Sulphur

Lapis lazuli and the ultramarines are blue coloured silicates ($Na_8Al_6Si_6O_{24}{\bullet}S_2$). The framework silicates are based on (Si, Al)O_2 frameworks with positive ions in the interstices. They also can contain negative S^{2-}. This negatively charged sulphur is thought to be affected by the aluminium polysilicates and sulphur radical species including S_3^- responsible for the blue colour are discussed[20]. The XPS properties of the ground material reveal the major sulphur portion to be in the oxidation state + VI. Only a very small amount could be assigned to a $\bullet S^-$, S^{2-} and/or S_3^- species at 161.3 eV.

It may appear rather strange to measure a strictly inorganic compound. However, there is one important correlation to biopolymers. In proteins the polypeptide chain imposes considerable strain on metal-sulphur binding centres which may lead to sulphur radical species. Such species have been discussed for cytochrome *c* oxidase[21,22] and plastocyanin[23]. Likewise the polysilicates could take over the same role. They are capable to polarize the sulphur species and the existence of sulphur radicals are debated.

In CuS the degree of covalency is substantially increased attributable to the interaction of Cu(II) with S^{2-}. According to our observations there is no Cu(II)S^{2-}. XPS

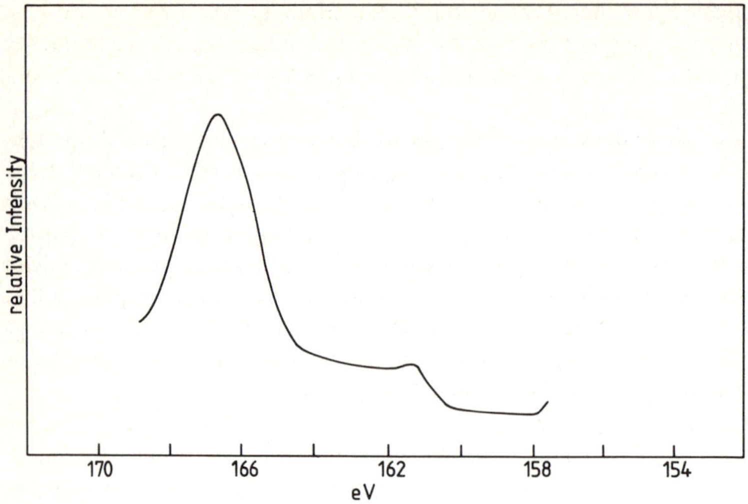

Fig. 1. Lapis lazuli: XPS spectrum of the S 2 $p_{(1/2,3/2)}$ region

measurements on CuS have shown one homogeneous signal at 932 eV which was assigned to 3 d^{10} Cu, the sulphur 2 s-signal was monitored at 161.1 eV. Unlike with CuO no satellites characteristic for Cu(II) were seen when CuS was measured (Table 1). Furthermore, no significant spectral changes compared to CuS were noted with NH_4CuS_4, an established source for Cu(I). Cu(II) satellites were also absent using $[Cu_{14}(D\text{-penicillamine})_{12}Cl]^{5-}$. This complex was chosen as it contains a thiolate-sulphur concentration very similar to that of the metallothioneins. Originally a mixed Cu(I)/Cu(II) valence complex was deduced from X-ray diffraction studies[23, 24]. Eight of the fourteen Cu are tetrahedrally arranged as usual for Cu(I). The remaining six Cu(II) were found to be in a square planar environment. The strong absorption in the visible region suggested some similarity with the type I copper chromophores. As S → Cu charge-transfer transitions do occur in both the Cu-penicillamine complex and the type I copper-

Table 1. Binding Energy Values of the Cu 2 $p_{3/2}$ and S 2 $p_{1/2,3/2}$ Levels

Compound	S 2 $p_{1/2,3/2}$	Cu 2 $p_{3/2}$	
		Satellite	Main peak
	eV	eV	eV
Cu_2O			932.5
CuO		943.8	934.6
NH_4CuS_4	162.1		931.5
CuS	162.1		932.3
Cu-penicillamine	163.2		933.3
Cu(8-mercaptoquinoline)$_2$	163.1	941.2	933.4
Cu(I)-Thiourea	162.2		932.5
Cu(I)-Thionein	162.0		932.3

proteins it was expected that the actual electronic state of the copper would be considerably affected. Regardless of the molecular architecture essentially all copper in the Cu-penicillamine is found in the $3d^{10}$ electronic state. According to Jørgensen the actual oxidation state was + 1[25].

It was intriguing to note the complete absence of characteristic Cu(II) satellites in the CuS spectrum. The objection might be raised that in general sulphur binding to $3d^9$ copper may enhance an initial photoreduction of the copper during exposure to X-ray irradiation leading always to $3d^{10}$ Cu. The ease of electron delocalization of the coordinated sulphur was thought to be the cause. When there is a second electron withdrawing ligand bound to the thiolate-sulphur, no S $\rightarrow$ Cu charge-transfer transition should take place. Indeed recording of Cu(8-mercaptoquinoline)$_2$ revealed a 2 $p_{3/2}$ band at 933.4 eV and the characteristic satellite structure attributable to $3d^9$ copper. Thus, photoreduction of Cu(II) coordinated by sulphur ligands can clearly be excluded.

4 Bovine Serum Albumin

Bovine serum albumin (M_r = 67000) is known to contain 4 methionine residues, 17 disulphide bridges and one cysteine residue. A metal driven Haber Weiss reaction, therefore, should be minimal and must be attributed to extraneously bond metal ions. One major band in the 163 eV S_{2p} region is seen (Fig. 2). According to the data bank published earlier[26, 27] overlapping signals each of methionine-sulphur, disulphide and the protonated cysteine-sulphur give rise to the appearance of this main signal. The second diminished S_{2p} binding energy near 168 eV may originate either from contaminating ammonium sulphate or sulphonic acid originating from one of these sulphur species. Of

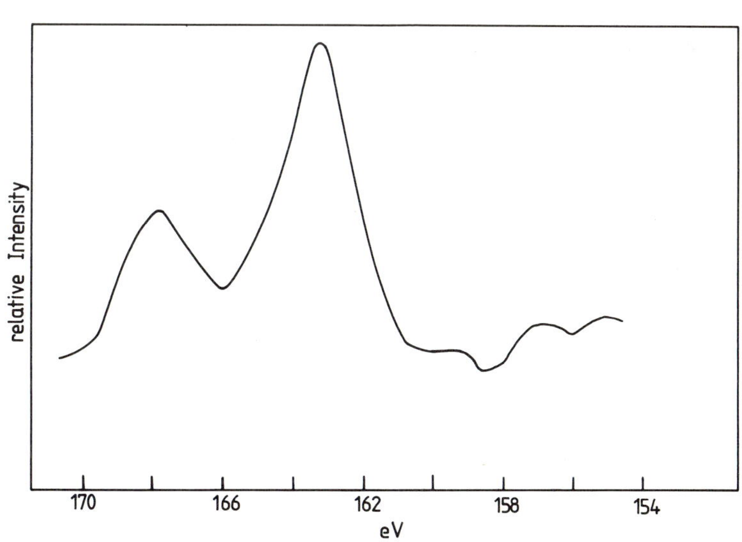

Fig. 2. Binding energy values of the S2p region of bovine serum albumin

course, XPS-spectroscopy cannot monitor one out of the 17 specific disulphide bridges or a certain methionine-sulphur out of four. The important conclusion is the general assignment of oxidised sulphur before or after a controlled redox reaction. Absolute assignments are only possible when one single sulphur containing amino acid residue is present. Prolongued irradiation at room temperature led to considerable oxidation of sulphur. Thus, all XPS-spectra were recorded at − 150 °C.

5 Papain

The XPS-spectrum of commercial papain desplays only a minor RSO_3^- and/or sulphate band above 168 eV. Essentially all sulphur has a 2p-binding energy level of 163.2 eV attributable to cystine and methionine sulphur (Fig. 3).

The power of XPS-spectroscopy must be seen in the fast and efficient control of the homogeneity of an isolated protein. Commercial samples are sometimes not homogeneous enough or tend to show age dependent deterioration. These can readily be seen by XPS. When rapid and thorough isolation of a protein can be accomplished, no oxidised sulphur species are seen. A good example proved to be Cd, Zn-thionein which had no active redox metals.

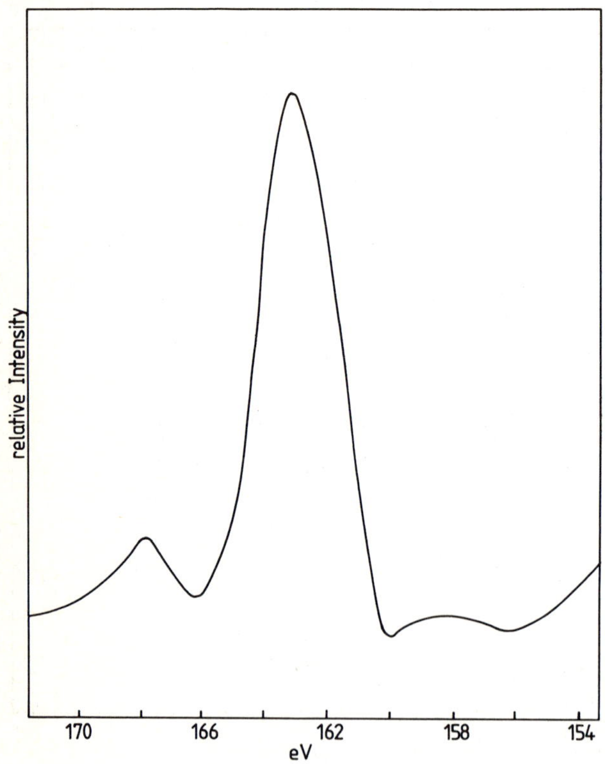

Fig. 3. XPS properties of papain. Recording of the S 2 p region

6 Cd, Zn-Thionein

The cysteine rich Zn, Cd-thionein displays one homogeneous thiolate sulphur signal at 161.7 eV. Virtually no higher oxidation state of the sulphur is seen. Upon treatment with excessive H_2O_2, sulphonic acid is the exclusive detectable sulphur species at 167.4 eV (Fig. 4).

When work was started on this protein only sulphur at 167.4 eV was seen[28]. Improved isolation techniques resulted in the recording of the above homogeneous thiolate sulphur signal[29].

7 Cu-Thionein

Unlike the Cd, Zn-thioneins, the respective copper-thiolate proteins were much more susceptible to oxidation. The binding energy value of the $S\,2_p$ core electrons of yeast copper thionein lies at 162.0 eV (Fig. 5). Ageing of protein samples or the addition of substoichiometric concentrations of H_2O_2 give rise to the appearence to transient oxidation states including RSSR, RSO^- and sulphonic acid. The copper thiolate

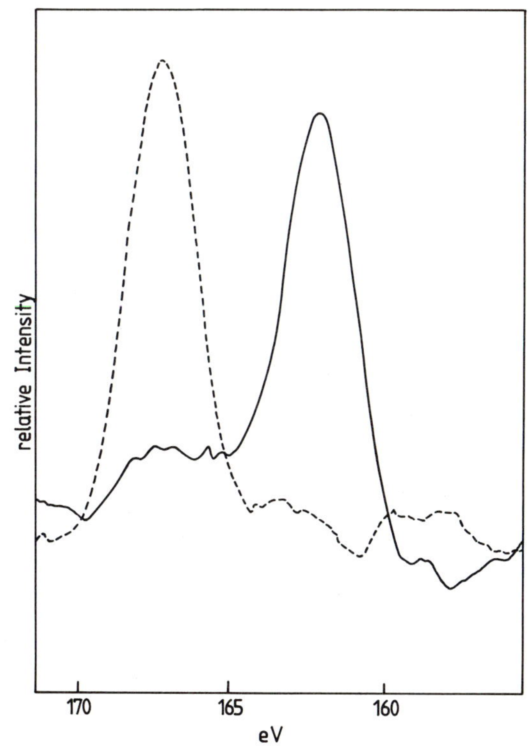

Fig. 4. XPS spectrum of deteriorated or aged preparation of rat liver Cd, Zn-thionein (----). Native Cd, Zn-thionein (—)

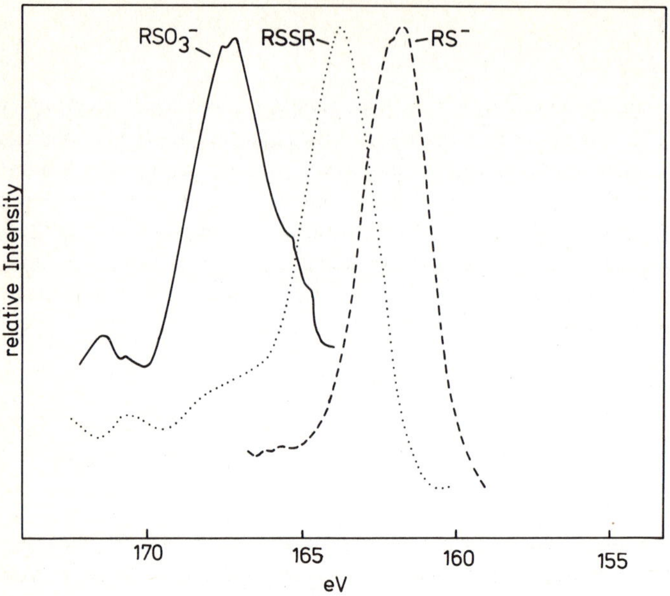

Fig. 5. S 2p binding energy values of fetal bovine liver Cu-thionein (----) native Cu-thionein (...) after 24 h (—) after 36 h. (with permission of Biochim. Biophys. Acta. *491*, 219 (1977))

chromophore[2, 14] is cleaved and irreversibly oxidised. In addition to the involved copper this oxidation is oxygen dependent (Fig. 5).

In comparing the Cu 2 $p_{3/2}$ and S 2p-values of Cu-thionein and Cu-thiourea, it was interesting to note virtually identical values. The double bonded sulphur of thiourea is probably able to withdraw electrons from the amide groups giving rise to this rather low S 2p binding energy value of 162.2 eV. In an earlier proposal of the structure of the Cu-chromophore[30–33], a tetrahedral arrangement of the sulphur atoms around the copper was assumed. The known crystal structure of Cu(I)-thiourea[14] encouraged this conclusion. (See also Table 1.)

8 Haem Proteins

The two differently bound methionine sulphur in the *b* and *c*-type haem proteins encouraged measurements of the respective S 2p binding energy value. In myoglobin, a representative of the b-type proteins, the methionine residue is exclusively located in the polypeptide side chain with no direct metal sulphur bonding. By way of contrast cytochrome *c* has one methionine coordinated to iron.

9 Myoglobin

Myoglobin from equine muscle contains sulphur exclusively in the form of methionine positioned at 55 and 131[34]. It was used to mimick a homogeneous thioether containing protein. Commercial samples showed S 2p binding energy values at 162.8, 165.5 and 166.8 with some contaminating sulphate above 168 eV. The lowest 2p value was assigned to methionine sulphur, those near 166.8 were attributed to sulphoxides probably originating from either radiation damage or iron dependent oxidation. After the addition of the oxidant Cu-phenanthroline the band at 162.8 eV was leveled off and the oxidised sulphur species including sulphoxides were more intense (Fig. 6).

10 Cytochrome *c*

Cytochrome *c* has 4 methionine residues, two of which are covalently linked to the haem moiety[35]. One of the other two methionine residues is coordinated to the iron in the axial position[36]. The major S 2p band of the crystalline compound appears at 162.6 eV attributable to the methionine residues. Prolongued irradiation causes an increase of the RSO_3^- or the sulphate band from 28% to 40% (Table 2). When aqueous cytochrome *c* is recorded, the amount of oxidised sulphur rises to 63% of the methionine sulphur band. The possible extraneously bound redox active transition metals, probably, have created a metal driven Haber Weiss reaction which led to the marked amount of oxidised sulphur observed. Splitting of the iron-sulphur bonding by cyanide results in dramatic increase of the 167.7 band and the additional appearance of a S 2p signal at 164.3, probably due to RS=O species. This oxidation is believed to be catalyzed by the haem iron. Hydrogen peroxide alone converts the methionine sulphur completly to sulphonic acid.

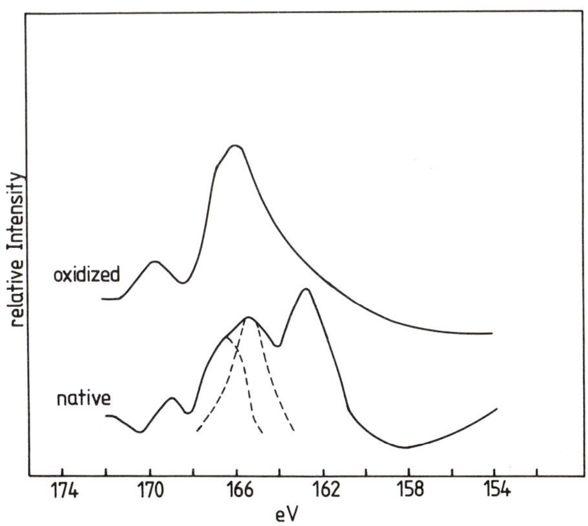

Fig. 6. S 2p binding energy levels of native and oxidised myoglobin

Table 2. Cytochrome *c*: Sulphur 2 p core electrons

Experimental conditions	Binding energy eV	Oxidised sulphur compared with methionine S 2 p values at 163 eV eV	(%)
Crystalline	162.6	166.1 (28)	
+ H_2O	163.0	166.0 (63)	
+ CN^-	162.4	164.3 (101)	167.7 (276)
+ H_2O_2	–	–	167.5 (222)
Crystalline after 6 h irradiation	163.3	169.1 (40)	

11 Plastocyanin

The two sulphur species originating from cysteine and methionine are involved in the copper coordination of plastocyanin. The sulphur 2 p binding energy values at 163.1 eV is highest in the reduced protein (Fig. 7). Oxidised sulphur at 166 is already detectable to a substantial degree. It is suggested that this binding energy signal is neither attributable to a S^+-Cu-coordination[37, 38], nor to extraneous inorganic sulphur[39] of higher oxidation state, but to a destruction of the protein. As in the case of Cu-thionein a metal driven

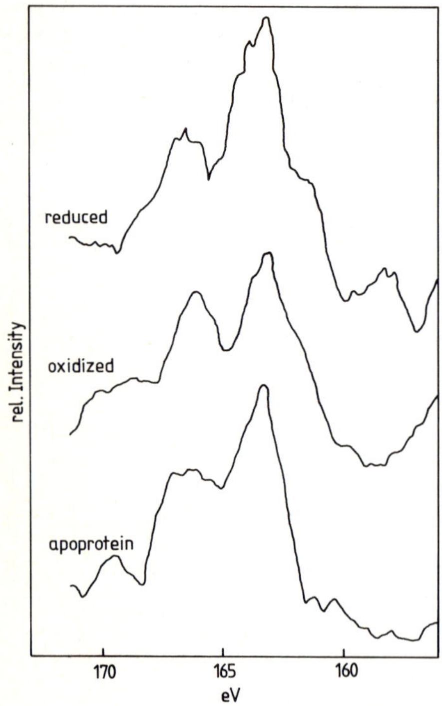

Fig. 7. XPS spectrum of apo-plastocyanin the Cu(II)- and Cu(I)-plastocyanin from spinach

Haber Weiss reaction has to be considered. In addition, a shoulder appears at 161.3 eV attributable to the Cu(I)-thiolate bonding. This shoulder is not seen in the oxidised protein, although it might be hidden under the relative broad ascending right side. The similar intensity of either sulphur 2p band at 166.3 and 163.1 should be noticed. The 163.1 eV band of the apoprotein rises sharply suggesting the complete dissappearance of any 161.3 eV contribution. The magnitude of the 167 eV band is still only half to that of the reduced holoprotein.

12 Cu_2Zn_2-Superoxide Dismutase

The first biochemical sample to be examined by XPS-spectroscopy in the author's laboratory was Cu_2Zn_2-superoxide dismutase (SOD). In the early days of isolating this copper- and zinc-protein, a second sulphur signal near 168 eV was always observed, which could not be fully explained[39]. As no ammonium sulphate was used in the course of the isolation, it was suggested that this S 2p signal may be attributed to a sulphur oxidation, originating from extraneously bound redox metals[2] usually found as traces in the employed buffer solutions. Furthermore, denatured haem-proteins, or fragments of Cu_2Zn_2-superoxide dismutase, were also thought to have contributed to catalyse the oxidation of sulphur containing amino acid residues. When the preparation technique was improved and under control, only one homogeneous signal at 163 eV was seen.

Excessive storage of the freeze dried SOD for several months led to the reappearance of oxidised sulphur.

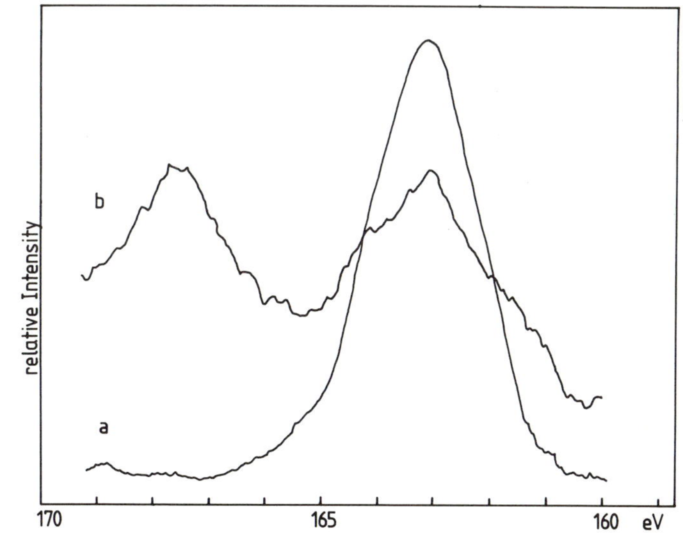

Fig. 8. S 2p levels of native (*a*) and deteriorated (*b*) Cu_2-Zn_2-superoxide dismutase

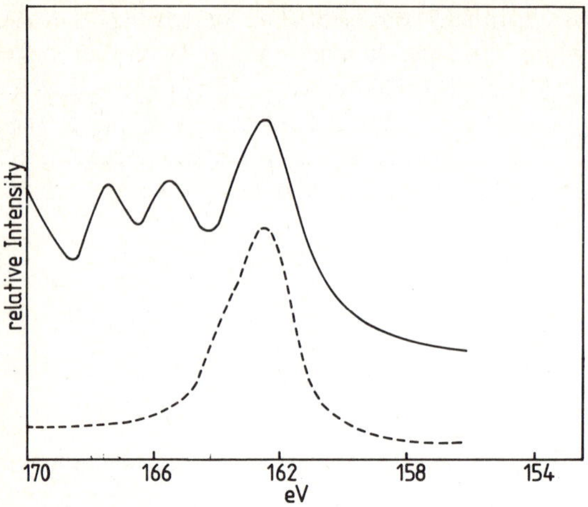

Fig. 9. XPS spectra of Cu_2Zn_2-superoxide dismutase isolated from red blood cells obtained from old (—) and young (---) bovine species

Considerable efforts have been made to use Cu_2Zn_2-SOD as a marker enzyme in studies of the biochemistry of ageing[40–43]. Usually there is a decline in the biochemical reactivity when Cu_2Zn_2-SOD from old species is compared with that of younger tissues. Concomitant with the diminished enzymic activity, it was of interest to examine whether or not the sulphur containing amino acids are also affected during the course of ageing. Cu_2Zn_2-SOD was isolated from red blood cells of old bulls using the strictly aqueous preparation technique and compared with the SOD obtained from erythrocytes of young animals (Fig. 9). Indeed, several sulphur 2 p-signals of the "old" SOD were monitored at 163.4, 165.4 and 168 eV, indicating a considerable oxidation of the sulphur containing amino acid residues. Only a slight shoulder at 164 eV was noticed when the "young" SOD was measured.

13 Selenium Containing Amino Acids

As in the case of sulphur, XPS was most suitable for studying electron transport on protein bound selenium, especially in those reactions where no magnetic properties or electronic transitions can be measured[26, 27, 39, 44–46]. The chemical shift of the binding energy values of the Se 3 d levels can reach up to 7.5 eV (Table 3). This large shift allows the assignment of the chemical nature of selenium and selenopeptides to be made[8, 9, 26]. Measurements of selenotrisulphides, the well known reaction products between selenite and thiols, deserved special attention (Table 4).

All selenotrisulphides show Se 3 $d_{(3/2,5/2)}$-signals at 55.5 ± 0.1 eV which are close to the values for elemental selenium and selenium methionine (Table 4). The earlier disputes dealing with the formal oxidation state of selenium in selenotrisulphides, ranging from 2+ to 2−, are now settled[47]. The actual oxidation state of Se lies near 1+.

Redox reactions of selenium could now be measured. A convenient example was the oxidation of selenium bound to sulphur in N-acetylcysteine-selenotrisulphide. Upon

Table 3. Electron binding energies of the Se 3 $d_{3/2,5/2}$ levels of some inorganic seleno compounds and of selenotrisulphides

Inorganic selenium compound	Se 3 $d_{3/2,5/2}$ eV	q (effective charge)	Oxidation number
PbSe	53.3	−0.09	−2
SnSe	54.2	−0.08	−2
As_2Se_3	54.5	−0.09	−2
Se	54.8	0	0
$Na_2SeS_4O_6$	56.1	0	+1
Na_2SeO_3	58.3	+1.04	+4
SeO_2	59.1	+1.04	+4
Na_2SeO_4	60.8	+1.31	+6
Selenotrisulphide			
Selenium di-mercapto-ethanol	55.4	0	+2/3
Selenium di-dodecane-thiol	55.4, 58.3	0	+2/3
Selenium di-N-acetylcysteine	55.5	0	+2/3
Selenium di-penicillamine	55.6	0	+2/3

increasing additions of H_2O_2, the band at 55.5 eV was progressively diminished and a second signal at 59.0 eV appeared. The signal was assigned to Se IV (Fig. 10).

With this data bank on selenium redox systems, it was possible to study the reaction mechanism of the selenoenzyme glutathione peroxidase[9]. Recording of the lyophilized protein revealed Se 3 $d_{(3/2,5/2)}$ signals at 54.4 eV. Stepwise additions of the natural acceptor substrate H_2O_2 shifted this signal to 58.0 eV, indicative to Se IV (Fig. 11). This signal was reversed following the addition of the donor substrate glutathione. From this observed chemical shift it was suggested that the enzyme bound selenium shuttles between selenol, or a selenol derivative, in its reduced form and in a seleninyl or selenonyl compound in its oxidised form. Later X-ray diffraction studies on the crystalline enzyme confirmed the presence of a selenocysteine in the active centre[48].

Table 4. Electron binding energies of the Se 3 d electrons in the different chemical environments

Compound	Chemical group	Se 3 $d_{3/2,5/2}$ eV
Selenium methionine	C–Se–C	55.1
Selenium cysteine	C–Se–Se–C	55.2
Selenium cystine	C–Se–S–C	55.3
Selenium di-penicillamine	C–S–Se–S–C	55.6
Sodium seleno-pentathionate	SO_3–S–Se–S–SO_3^{2-}	56.1

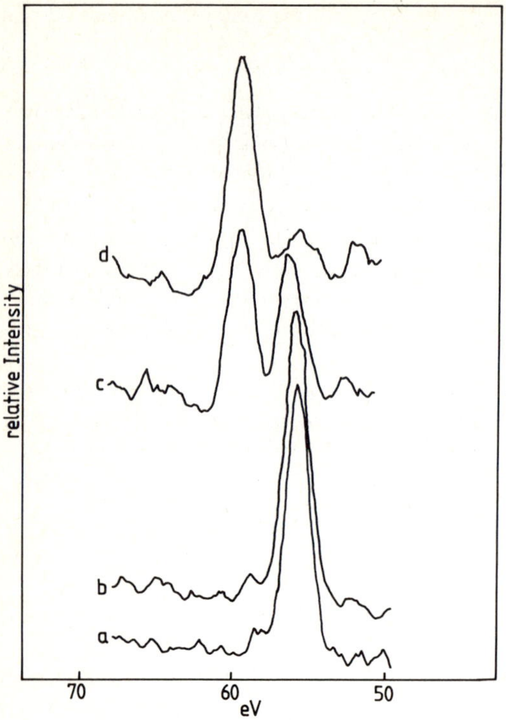

Fig. 10. XPS spectra of Se 3 $d_{(3/2,5/2)}$ electron binding energies employing Se-di-N-acetylcysteine titrated with H_2O_2. (*a*), (*b*) substoichiometric amounts of H_2O_2, (*c*), (*d*) excessive H_2O_2-concentrations (with permission of J. electron spectroscopy and related phenomena *10*, 434 (1977))

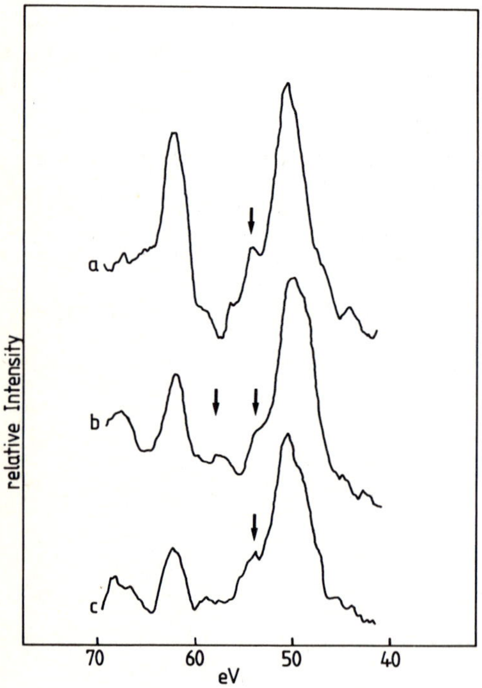

Fig. 11. XPS spectra of glutathione peroxidase. The arrows are indicative for Se 3 $d_{(3/2,5/2)}$ lying between Na 2 s at 62.2 eV and Mg 2 p at 50.8 eV. (*a*) The native enzyme, (*b*) after the addition of H_2O_2, (*c*) reduction of the H_2O_2 oxidised enzyme with glutatione (with permission of Biochim. Biophys. Acta *277*, 213 (1975))

14 Conclusion

XPS-spectroscopy of sulphur and selenium containing amino acid residues in biopolymers has proven to be a valuable tool for examining redox reactions of these chalcogens. In contrast to laborious chemical approaches, where substantial modifications of the involved proteins are required, X-ray photoelectron spectroscopy is both fast and efficient. The kinetic energy of the emitted photoelectrons is high enough to allow the detection of a single sulphur or selenium per 20 000 Daltons. Furthermore, the use of this technique allows a quick decision on the rate of deterioration of a protein. The limitations must be seen in that the oxidised chalcogens cannot always be assigned to a specific amino acid residue.

Acknowledgements. The many experimental sections were supported by grants from the "Deutsche Forschungsgemeinschaft". Part of this review was written at the Istituto di Chimica Biologica, Università di Roma, Italy. The travel support by the "Thyssen Stiftung" is gratefully acknowledged. Special thanks go to Prof. Dr. G. Rotilio for his helpful comments.

Abbreviations

XPS = X-ray photoelectron spectroscopy
SOD = Superoxide dismutase, superoxide: superoxide oxidoreductase (EC. 1.15.1.1.)

15 References

1. Weser, U., Hartmann, H. J.: Copper thiolate proteins (metallothionein), Copper proteins and copper enzymes, Vol. 3, p. 151, Boca Raton, CRC press 1984
2. Hartmann, H. J., Weser, U.: Biochim. Biophys. Acta *491,* 211 (1977)
3. Metallothionein, Proceedings of the "First International Meeting on Metallothionein and other Low Molecular weight Metal-binding Proteins" (Kägi, J. H. R., Nordberg, M., eds.), Basel, Birkhäuser Verlag 1979
4. Hartmann, H. J., Morpurgo, L., Desideri, A., Rotilio, G., Weser, U.: FEBS Lett. *152,* 94 (1983)
5. Morpurgo, L., Hartmann, H. J., Desideri, A., Weser, U., Rotilio, G.: Biochem. J. *211,* 515 (1983)
6. Haber, F., Weiss, J.: Proc. Roy. Soc. Ser. A *147,* 323 (1934)
7. Pilz, W.: Dissertation, Tübingen 1978
8. Weser, U., Sokolowski, G., Pilz, W.: J. Electron Spectr. *10,* 429 (1977)
9. Wendel, A., Pilz, W., Ladenstein, R., Sawatzki, G., Weser, U.: Biochim. Biophys. Acta *377,* 211 (1975)
10. Rupp, H., Weser, U.: ibid. *446,* 151 (1976)
11. Jørgensen, C. K., Berthou, H.: Photo-electron Spectra Induced by X-rays of Above 600 Non-metallic Compounds Containing 77 Elements, Mat. fys. Medd. Dan. Vid. Selskab (Copenhagen) *38,* no. 15 (1972)
12. Adams, I., Bancroft, G. M.: Nature (London) *250,* 219 (1974)
13. Prinz, R., Weser, U.: Z. Physiol. Chem. *356,* 767 (1975)

14. Weser, U., Hartmann, H. J., Fretzdorff, A., Strobel, G. J.: Biochim. Biophys. Acta *493,* 465 (1977)
15. Bordas, J., Koch, M. H. J., Hartmann, H. J., Weser, U.: Inorg. Chim. Acta *78,* 113 (1983)
16. Sokolowski, G., Weser, U.: Z. Physiol. Chem. *356,* 1715 (1975)
17. Gärtner, A., Weser, U.: FEBS Lett. *155,* 15 (1983)
18. Brütsch, S., Hartmann, H. J., Weser, U.: Inorg. Chim. Acta *92,* 147 (1984)
19. Clark, R. J. H., Dines, T. J., Kurmoo, M.: Inorg. Chem. *22,* 2766 (1983)
20. Seel, F., Schäfer, G., Güttler, H. J., Simon, G.: Chemie in unserer Zeit *8,* 65 (1974)
21. Brunori, M., Antonini, E., Wilson, M. T.: Metal ions in biological systems (ed. Sigel, H.), Vol. 13, p. 187, New York, Basel, Marcel Dekker, Inc. 1981
22. Malmström, B. G.: Biochim. Biophys. Acta *549,* 281 (1979)
23. Weser, U., Younes, M., Hartmann, H. J., Zienau, S.: FEBS Lett. *97,* 311 (1979)
24. Birker, P. J. M. W. L., Freeman, H. C.: J. Am. Chem. Soc. *99,* 6890 (1977)
25. Jørgensen, C. K.: Adv. Quantum Chem. (Löwdin, P. O., ed.), Vol. 8, p. 137, New York, Academic Press 1974
26. Rupp, H., Weser, U.: Bioinorg. Chem. *5,* 21 (1975)
27. Sokolowski, G., Weser, U., Pilz, W.: FEBS Lett. *48,* 222 (1974)
28. Simon, A.: Biochemical master thesis, Tübingen 1972
29. Weser, U., Rupp, H., Donay, F., Linnemann, F., Voelter, W., Voetsch, W., Jung, G.: Eur. J. Biochem. *39,* 127 (1973)
30. Rupp, H., Weser, U.: Biochim. Biophys. Acta *533,* 209 (1978)
31. Weser, U., Rupp, H.: The Chemistry, Biochemistry and Biology of Cadmium, p. 267, Amsterdam, Elsevier/North Holland Biomedical Press 1979
32. Weser, U., Schubotz, L. M., Younes, M.: Copper in the Environment, Part II: Health Effects (ed. Nriagu, J. O.), p. 197, New York, Wiley Interscience 1979
33. Weser, U., Rupp, H.: Metallothionein, Experientia Supplementum 34, (eds. Kägi, J. H. R., Nordberg, M.) p. 221, Basel, Birkhäuser Verlag 1979
34. Bradshaw, R. A., Gurd, F. R. N.: J. Biol. Chem. *244,* 2167 (1969)
35. Margoliash, E., Schejter, A.: Adv. Protein Chem., Vol. 21, p. 113, New York, Academic Press 1966
36. Dickerson, E., Takano, T., Eisenberg, D., Kallal, O. B., Samson, L., Copper, A., Margoliash, E.: J. Biol. Chem. *246,* 1511 (1971)
37. Solomon, E. J., Clendering, P. J., Gray, H. B., Grunthaner, F. J.: J. Am. Chem. Soc. *97,* 3878 (1975)
38. Wurzbach, J. A., Grunthaner, F. J., Dooley, D. M., Gray, H. B., Grunthaner, P. J., Gay, R. R., Solomon, E. J.: ibid. *99,* 1025 (1977)
39. Jung, G., Ottnad, M., Bohnenkamp, W., Bremser, W., Weser, U.: Biochim. Biophys. Acta *295,* 77 (1973)
40. Bartosz, G., Soszynski, M., Retelewska, W.: Mech. Ageing Dev. *17,* 237 (1981)
41. Bartkowiak, A., Grzelinska, E., Bartosz, G.: Int. J. Biochem. *15,* 763 (1983)
42. Gershon, D.: Mech. Ageing Dev. *9,* 189 (1979)
43. Glass, G. A., Gershon, D.: Biochem. Biophys. Res. Comm. *103,* 1245 (1981)
44. Prinz, R., Weser, U.: FEBS Lett. *54,* 224 (1975)
45. Jung, G., Ottnad, M., Bohnenkamp, W., Weser, U.: ibid. *25,* 346 (1972)
46. Weser, U., Donay, F., Rupp, H.: ibid. *32,* 171 (1973)
47. Martin, J. L.: Organic Selenium Compounds: Their Chemistry and Biology (eds. Klayman, D. L., Günther, W. H. H.), New York, John Wiley 1973
48. Ladenstein, R., Epp, O., Bartels, K., Jones, A., Huber, R., Wendel, A.: J. Mol. Biol. *134,* 199 (1979)

Author-Index Volumes 1–61